The Sea Chest

A YACHTSMAN'S READER

THE

Sea Chest

A YACHTSMAN'S READER

Edited by Critchell Rimington
FORMER EDITOR AND PUBLISHER OF Yachting

W · W · NORTON & COMPANY · INC · *New York*

Library of Congress Cataloging in Publication Data
Rimington, Critchell, 1907–   ed.
    The sea chest.
    Stories and articles selected from the Sea chest,
the yachtsman's digest.
    1. Yachts and yachting—Addresses, essays, lectures.
2. Seafaring life. I. The Sea chest.
GV813.R5   1975   797.1′08   74–28144
ISBN 0–393–03183–7

1 2 3 4 5 6 7 8 9 0

TO
JAY

Contents

Introduction to the New Edition

FOR A period of about five years prior to the last World War it was my pleasure to publish and edit a yachting digest which achieved a modest reputation under the title of *The Sea Chest: The Yachtsman's Digest*. This capsule publication came into being for the purpose of preserving a selection of those articles and stories relating to the yachting scene which, by reason of their excellence, gave every indication of achieving something more than transitory recognition.

This volume, now reprinted, represents a winnowing of the various issues of *The Sea Chest*. The selection is based, it must be confessed, on the editor's personal preference. Hence it is not to be said or even implied that the pages which follow represent anything more than a collection of those many stories and articles which have brought endless pleasure and edification to the countless yachtsmen who have read them in years past. Yet it is both my hope and real belief that in permanent format they will provide added enjoyment for their old friends as well as for a new generation which will discover them for the first time.

I am particularly happy to again have this medium of expressing my sincere appreciation to the many authors, editors and publishers whose generous co-operation have made this volume possible.

C. R.

Rowayton, Conn.

To Bermuda in "Wee One"

HARRY A. YOUNG

FROM the day I first gazed upon *Wee One's* dignified completeness, I knew that here was the seagoing small boat I had always wanted. True, she was smaller than I had planned but her record showed her to be far above the average boat of her size under sail, and one had but to look to see seaworthiness written all over her.

Her wide decks, deep self-bailing cockpit, short ends, and apparently excessive freeboard (which, by the way, I would not reduce one inch) make up a sturdy little vessel, as dry and well mannered in any sort of going as one could wish for. Besides her ability in rough water, her lines and rig have produced a fast little boat and one that will go to windward even in a nasty chop. On an over-all length of 20 feet, Fred Geiger, her designer, gave her 6 feet 4 inches beam and 4 feet draft. Within these abbreviated dimensions, *Wee One* has 5 feet 6 inches headroom, two berths, toilet, galley, and a two-horsepower Palmer engine which is out of sight in its own compartment under the bridge deck. She is rigged as a jib-headed cutter and spreads a total of 303 square feet in her working sails.

From the day she became mine to the day of our departure for Bermuda, my spare time was spent brushing up on my self-taught navigation, checking the rate of a watch which I had set in gimbals, and puzzling over the water storage problem. *Wee One* had a 10-gallon water tank aft. To this we added a 15-gallon temporary tank forward and stowed another 15 gallons in one-gallon tins in the after ends of the sail lockers which extend under the deck to the transom alongside the cockpit. In this way we could keep her in proper trim by drawing evenly fore and aft on our water supply. Forty gallons of water was

Reprinted by permission of, and copyright 1938, by *Yachting*.

thought to be sufficient to see us back to the coast if I could not locate the islands.

To prepare properly for an off-shore voyage in one's own boat requires an enormous amount of forethought and hard work. There are so many different situations and weather conditions that may be met, to say nothing of the provision problem, that I do not think I would have been able to manage without the capable assistance of my mate, Roy Disney.

We left our slip near Baltimore at noon on Saturday, May 29th, both of us fearful that something would happen at the last moment to prevent the trip. Sunday morning, the 30th, found us a little more than half way down the Chesapeake. The afternoon brought forth our prevailing southeast wind, giving us a dead beat the rest of the way. Rather than buck this and a flood tide, we put into Harborton, Virginia, to have some one show us why the engine refused to run. The morning of the 31st was spent making repairs on the motor and the afternoon saw us on our way again in the face of a double reef breeze out of the south— not much better than the previous day.

During the following twelve hours or so of tiresome beating, we remarked more than once upon our uncanny ability for finding head winds on our vacation trips. At 11:30 a.m., June 1st, we took our departure from Cape Charles, setting a course slightly to the north of Chesapeake Lightship. The wind, now southwest and abeam, sent us along at five knots, under main, staysail and genoa. Our friends, the porpoises, playfully fought for the privilege of swimming under the bowsprit and at last we were at sea with a fair breeze.

The first day's run gave us 98 miles on our way and placed us in the blue water of the Gulf Stream. Light winds were the order of the second day and a swim in the delightfully warm and crystal clear waters of the Stream was enjoyed immensely. One of us stood on watch for sharks while the other went overboard.

The southwest wind slowly increased during the afternoon and by evening had us down to double reefed main and staysail. We hove to under staysail at 11:00 p.m., wanting to see how she acted in the rollers which were increasing in size by the minute. My decision to heave to was due to a determination to have no mishap at this time. We had learned to judge our speed through the water rather accurately but wind strength is a different story. We supposed at the time that it was blowing 30 or 35 miles an hour, but between 25 and 30 is probably more nearly correct.

Heaving *Wee One* to is done by furling everything except the staysail. The sheet of this sail is next eased off slightly. The tiller on this occasion was lashed hard down to leeward. It was found that she gradually worked up to windward until the staysail luffed, after which she would quickly fall off until wind and sea were abeam. Hesitating here for a moment, she would start the cycle over again. During this cycle probably ten waves passed under her. Later in the cruise we found that we could adjust the tiller to slow up this process and luff the sail completely only once in a great while. Adjusted so, she lay between five and six points off the wind and made about as much leeway as headway. This, I believe, is fairly accurate because I have successfully figured my course while hove to as at right angles to the wind. Under these conditions, she makes about a half knot through the water. After observing her actions for a short while, my uneasiness gave way to increasing confidence. Our motion was much easier in the large swells than it had been in the short Chesapeake chop.

Hove to thus, we were both able to go below and sleep soundly, doubled in a knot, our knees jammed against the square mast and our backs against the sides of the ship. We found the mast location between the bunks useful for this purpose. Morning found the wind reduced, probably to twenty knots. The seas, however, seemed enormous to us. Wave heights, I know, are as deceptive as wind strength. Our mast, however, is approximately 35 feet above the water line and, when in a trough between two large ones, the top seemed about even with the crests. She seemed comfortable under a double reefed main, staysail and flying jib, easing herself gracefully over the crests, wind abeam and sheets eased.

About 6:00 p.m. that afternoon, I was surprised to see, out of the corner of my eye, a towering dark object astern. It was the tanker *Gulfwing,* her skipper on the bridge, megaphone in hand, ready to come alongside. His hail was: "Are you all right?" On being told that we were, he asked if he could be of any help. A position was requested which he gladly gave with his best wishes for a successful trip in what he called a mighty small boat.

As soon as the position was plotted and we saw how closely it checked with mine, we decided to celebrate the event by opening a can of baked chicken for dinner. While on the subject of food, I might mention that our diet lacked nothing on this first half of the voyage. Canned beef stew, lamb stew, various soups, corned beef hash ready to fry, powdered milk (a good substitute for the real thing), tomato juice, grapefruit juice, fruit cocktail, chicken, etc. A large can of fruit cocktail divided between the two of us made a delightful lunch each day and no doubt had much to do with keeping us fit. Eggs kept well in a container in the bilge and the butter, lard, etc., were placed under the bunks, where they were fairly cool. Ice, of course, was impossible. The ice box, from start to finish, was used for stowing things other than ice.

The following morning, the steamship *Luossa,* of Stockholm, repeated the performance of the *Gulfwing.* We did not ask our position this time, however. Neither skipper seemed to think we belonged so far offshore, which was natural enough, I suppose, since *Wee One* is shorter than their lifeboats! During the day we had the misfortune of having the tank of our pressure stove burst and spill a half gallon of denatured alcohol over the bunks, clothing, through the lockers, and into the bilge. We were extremely fortunate that this did not ignite but a more sickening mess I can't imagine. The odor would be most unpleasant even ashore. In the sea that was running, five minutes spent below was enough to make us rush for the deck, praying for fresh air. A primus, set in gimbals, has long been my favorite stove and this experience clinches its place in my preference.

The following five days gave us delightful breezes, ranging from south to west, strength, 10 or 15 miles, and mostly abeam. *Wee One* loves this going and with her beautiful genoa, staysail and full main drawing nicely, we ate up the miles in the moderate sea that was running. Steering her thus is a positive joy, and only occasionally is a little pressure on the tiller required. The noise from her bow is like nothing so much as someone below, snoring.

Our day's run was usually around 80 or 90 miles. One day, however, we were able, by running slightly off the course, to set the spinnaker for 24 hours. This day, the best of the trip, put 120 miles behind us.

Midnight of June 9th found us expecting to sight Gibbs Hill and St. Davids Light to the south within three or four hours. We had approached from the northwest and planned to clear North Rock by about ten miles. This would have us 20 miles offshore but still within range of the lights in clear weather. However, as we learned later, thunder squalls played around the islands all evening and early morning. We apparently came abreast about 10:00 p.m. because at 11:00 we detected the scent of flowers and must have been down to leeward of the lights. Flies had not bothered us for more than a week but next morning there were two on deck.

A strong easterly set must have accounted for our premature arrival. I later learned from "Roddy" Williams, of the Royal Bermuda Yacht Club, that strong easterly currents around the island attend the prevailing southwest winds. A cloudy morning made a longitude sight impossible and, in my excitement, I forgot to clamp the sextant arm for the noon sight. Roy can certify that during the afternoon I just about wore the figures off the sextant scale. Sights were numerous, to say the least. By 1:30 p.m. I had fairly well established our position as 50 miles ENE of St. Davids. Our course was immediately altered to W by S. About 5:00 p.m., a warship, the first object, other than an empty drum and a gas tin, sighted in five days, showed up ahead and several hours later numerous flies came aboard. These we took as a good sign that land was ahead as our course was close hauled.

At midnight, St. Davids Light was sighted on the port bow and the reflection of Gibbs Hill in the sky on the starboard bow. We both stayed awake all night to celebrate; I, chiefly because my navigation was justified, and Roy because he had only three days left to catch a liner which would get him home in time for his job. Light head winds and a cranky engine delayed our arrival at St. Georges until noon, twelve hours after sighting the lights.

We hoisted our quarantine flag at Five Fathom Hole but arrived at the dock unnoticed. After examination by the port doctor, we were directed to a mooring alongside the town wharf where Arthur Woodman, genial host of the White Horse Tavern, took over *Wee One* and her crew. He did a good job of making our stay in Bermuda pleasant.

The fourth day after our arrival was spent signing on a crew and provisioning the boat for the voyage home. Lionel McCallan, a native Bermudian who had spent much time on small seagoing yachts, made an excellent crew.

A thorough study of weather conditions over our route for the past fifty years showed a prevailing SW wind, hurricanes unknown and a *small* percentage of calms for the month of June. We could, therefore, reasonably expect good sailing conditions for the voyage home, which was to be either by way of the Chesapeake or Delaware Bay. The actual mileage varies slightly and the final course was to be determined by wind direction on approaching the coast. The trip down, with an unfaltering SW wind, had further beguiled me into expecting the same conditions for the return.

Early on June 15th, after a tow out of St. Georges Harbor by Mr. Stubbs' *Princess,* full sail was set for the return. The wind was light northerly and lasted just long enough to take us out of sight of Gibbs Hill Light, after which a calm set in lasting four monotonous days and nights. We had been unable to obtain repair parts for the engine in Bermuda, so nothing was to be done but wait for wind and try to take advantage of the faint zephyrs which occasionally rippled the surface of the glassy ground swell. Since neither of us is of a nervous disposition, the first several days were no hardship. At the end of the third day, however, we were slightly less than 70 miles from Bermuda and beginning to worry.

Morning of the fifth day brought us a whole-sail SW wind which allowed us to slack sheets and really move. By evening, we grudgingly turned a double reef in the main. When the increasing wind forced us to heave to under staysail early next morning, we cursed our luck. Twenty-four hours later, the staysail, which was hammering badly each time a wave threw her into the wind, was furled and we ran before it under bare poles. We decided that running before it would save the sail and rigging and would help us a little on the course. Until 3:00 p.m. the following afternoon, wide-awake steering was imperative as about every second sea was breaking, each of them a menace.

The center of the depression (reached in mid-afternoon of the 22nd) was attended by intense rain and wind. My guess about the wind force is of little value, and we had no instruments for this purpose. However, water was being whipped from the crests in a straight line, making the rain taste salty. By 6:00 p.m., the wind had moderated considerably and by dark we hove her to and both went below for rest and a chance to remove drenched oilskins.

Next morning gave us a change of wind direction as well as a considerable decrease in velocity. The resulting cross sea made the going poor under double reefed main, staysail and jib. Although we were anxious to ascertain our position, a sight was impossible in this sea and I gave up after a few attempts. During this time, however, I did locate the best position from which to take a sight in bad weather. By lowering and furling the mainsail in its boom crotch, I was able to straddle the boom, facing aft, my toes under the forward edge of the companion slide, much as a man rides a horse. This position left my hands free for the sextant.

Then followed three days of light winds during which time we crossed the Gulf Stream. Due to wind direction, a final course was laid for Delaware Bay and at 4:30 a.m., June 28th, the occulting light of Five Fathom Bank Lightship was located abeam.

The morning's run through a dense fog placed us within hearing of the horns on Overfalls Lightship and those of Breakwater and Harbor of Refuge Lighthouses at the mouth of Delaware Bay. This is a nasty piece of water and we were thankful for a shower which conveniently chased away the fog but, unfortunately, left us becalmed. Jack Hill, of the Emergency Lighthouse Service, happened along at this time and kindly offered a tow into Lewes, Delaware, where we dropped anchor at 11:00 a.m., June 28th. Both Mr. Hill and Captain Derickson, of the Coast Guard at Lewes, were most kind. The latter, after considerable tinkering, managed to make our engine function once more. The remainder of our voyage was uneventful and we arrived at our home anchorage in Jones Creek at 2:00 a.m., Thursday, July 1st.

I have always been of the opinion that Captain Tom Day knew what he was talking about when he stated, and later proved with *Sea Bird,* that size is not particularly important for offshore work provided your ship is well designed, built and equipped. *Wee One* is probably one of the smallest boats to make a passage of this kind successfully, but I can truthfully say that at no time did she give me any anxiety as to her ability.

⚓

2

Down Spinnaker!

PHILIP HOLLAND

WE WERE late getting aboard, which was a pity, because George always has to allow himself time to hoist the mainsail and then lower it again to put the battens in. So we were in a great hurry, and the five-minute gun found us with only the foresail set, and the *Mistral* still on her moorings. It was not an uncommon situation for George and myself; we solve it by letting the foresail draw and dropping the mooring over the lee side, if the wind is offshore, so that they can't see us from the club-house. Once under way, nothing very much happened for the next few minutes, so I'd better tell you something about George, so that you'll appreciate the extraordinary events of that race.

George has only once won a race; before the time I'm talking about, that is. That was when all the boats except *Mistral* and *Jeunesse* started on the wrong gun, and *Jeunesse* fouled a mark, so George couldn't help winning. George is one of those chaps who seem to have the average amount of bad luck, but lack the compensating dollop of good luck. If he ever looks like winning a race, something always goes wrong—and when his chances of winning look pretty remote, as they usually do, nothing ever happens to give him that helping hand that other helmsmen sometimes get. Nothing, that is, until the Smythe Cup Race last year.

We were, as I have said, late in getting under way, and for the last few minutes before the gun we were so busy getting things stowed down and set up that the starting-gun found us way down at the leeward end of the line, and some fifty yards away from it. And on the wrong tack. However, this was not unusual, and we were soon pursuing our placid way at the stern of the fleet, untroubled by rights of way or over-

Reprinted by permission of *The Yachtsman*.

taking rules. There was a nice whole-sail breeze, and the sun shone warmly on our necks.

George had suggested reefing about half a minute before the start, but I persuaded him against the idea, although, as he said, we should probably get our transoms beastly wet. I've always crewed for George. I suppose the reason is partly that no one else will do the job, and partly that without me George would probably never get started at all. Furthermore, we are temperamentally suited. George realizes that if he should want to come about while I'm trying to light my pipe, he should defer the maneuver until the weed is drawing nicely. While, I, on the other hand, am always ready to sacrifice my handkerchief when George finds he has left the racing flag ashore.

The wind began to freshen as we beat towards the first mark, and by the time we'd rounded it, there was quite a lumpy sea running. The second leg was a reach, and we saw that the others were reefing, for, with the fickleness of summer winds, the breeze was now definitely too much for the whole sail. George, after asking me whether it would be convenient, luffed *Mistral,* and I started to roll down a reef with the patent gear. I'd just managed to get the handle of the rachet unstuck, when there came a sudden free puff. The mainsail caught the weight of it, and the handle snapped in my hand. Of course, there was a fault in the metal, but knowing that didn't help us much.

There was nothing for it but to carry on. We tried reefing the foresail alone, but that made her so unbalanced that she nearly pulled George's arm out. In the midst of our troubles we paused to look around us. We found, to our surprise, that we'd caught up the rest of the fleet, which was fairly bunched together. We also noticed that the race officer had hoisted a signal which, after much shouting between the boats, was interpreted to mean that the race was shortened to one round.

This was welcome news, for by now several of the boats had reefed again. They argued, wisely, that it was better to lose a race than a mast. For George and me things were more hectic every minute. George started the mainsheet so that the leach was cracking like a machine-gun. We were both lying out as far as we could, not because it was doing much good, but because we felt that it was. The situation was indeed desperate. Ahead of us lay a mark round which we must jibe. After that lay a dead run to the finishing line. With almost the whole mainsail set and the sea now running, it would be almost impossible to hold her, even if we didn't lose the mast. We discussed the situation, and noticed, meanwhile, that we were now fourth, with three boats astern of us.

Finally we decided that the only thing to do was to cut the main halliard and run home under the foresail. Of course, there were other things we could have done, but we didn't think of them till afterwards. It was then we found that we had no knife. It took some time to convince us, but finally we had to face up to it. No knife, nothing that would cut. So we decided to wear ship—but we couldn't. There were three boats close to weather of us, and shout as we might, we couldn't make them understand the situation. It was impossible to luff without hitting them, and traveling as we were, heaven knows what would have happened then.

Before I relate the amazing events of the next few minutes, I'll try and clarify the situation, so that you'll understand how it was that George won the Smythe Cup.

Our boats are half-decked C.B. sloops, 18 ft. l.o.a., and inclined to be a bit tender. On a run we set a parachute spinnaker, which is usually set in stops before the race, as it is not advisable to go for'ard when you're running. *Gypsy* ran under once, so now we have the spinnaker all ready, with the sheets leading aft, so that we can break it out when we want it.

On this memorable occasion, everyone had their kites in stops, in the usual way, only, of course, no one thought of setting them. In that sea, it would have been just damn silly. As we approached the mark, *Mistral* was still fourth, each of us overlapping the other, so that we had the inside berth at the mark. Just before we jibed, George asked me in a harassed voice to let the plate down, to help her round. My hands were cold, and the rope slipped, so that the plate went down with a run. A bight of the spinnaker's starboard sheet fouled the drum, and the parachute burst from its stops with a horrible "whoom-bang." At this moment we jibed. *Mistral* flung herself over like a car taking a bend on two wheels. The bilge-water whooshed across the floorboards as I grabbed the coaming, and hoisted myself up to windward, where I should have been in the first place.

The next few minutes are somewhat blurred. George was yelling like a dervish and trying to set up the port backstay, and let go the starboard one, all in one operation. At the same time he was shoving the tiller to weather with his stern, to stop her from broaching-to. He shouted to me to let the spinnaker halliard go. I fished around in the bilge and found the fall of the halliard, but when I cast it off, the sail came down about a foot, and then stopped. The harder I heaved, the tighter it jammed. I looked aloft, and saw that the spinnaker was on the wrong side of the forestay, which meant that the halyard had a foul lead round the jumper stay. If I let go the one movable sheet, the sail would flog itself to pieces, if it didn't bust the mast first.

So I left it, and came aft to help George. He was somewhat calmer now. In fact he was staring in amazement at the water rushing by. *Mistral* certainly was going like a train. *Josephine,* who had been leading, was now clear astern of us. In fact, if the mast hadn't been whipping about like a trout-rod, we should have been quite happy. The seas coming up astern had a nasty gleam in their eye, but I wouldn't let George look, and we weren't pooped. In fact, *Mistral* behaved in a most ladylike manner, although George kept swearing at her when she seemed inclined to wander from the straight and narrow course.

We won the race by twenty-three seconds, and two seconds after we had crossed the line the mast snapped. Which solved the problem of stopping, anyway.

⚓

3

The Taming of the Parachute

ERNEST RATSEY

IN ABOUT the year 1866, in England, the cutter *Sphinx* rounded the weather mark and, to her competitor's surprise, set a large, peculiar-looking sail on the opposite side of the hull to which she was carrying the main boom. For the want of a better name this sail was dubbed the "*Sphinx's* Acre," which nowadays has been shortened to "Sphinxer," or just plain "spinnaker." Prior to this date the squaresail was used for running, but I understand that the idea of the spinnaker was formed by seeing the Thames barges boom out to windward the clew of the large forestaysails when running. From that date until about 1922, spinnakers always had the same main characteristics—that is, they were narrow up at the head and varied in the width at the foot, depending upon the length of the spinnaker pole and whether the user wanted a large or small sail.

Generalizing, we can say that the older gaff-rigged yachts had low, broad sail plans, which gradually, year by year, became higher and narrower until the adoption of the Bermuda rig, when booms no longer overhung the counters and bowsprits for the most part were discarded.

Towards the end of this transition period, boats were handicapping themselves for running "down wind" by their mainsails becoming higher and narrower, and their spinnakers altering likewise, inasmuch as the maximum length allowed for the spinnaker pole was limited to the base of the fore triangle, for the larger yachts under the Universal Rule, and all classes, large and small, under the International Rule. The smaller yachts under the former rule had a more generous allowance, viz: 40% of their sail plan's base line.

This meant that running before the wind one had a large, square-shaped mainsail on one side and a long spinnaker pole sticking out on the other. Yacht owners seemed satisfied with these spinnakers that were narrow in the head, because owing to the long spinnaker poles, the sails themselves seemed to have ample area. Then, too, in the larger yachts spinnakers were not used if it was possible to set a balloon jib and tack down wind, first on one jibe and then on the other, for this was found to be faster than setting the spinnaker and running down before the wind. (The gaff mainsail is still recognized as a faster sail than a marconi sail for reaching.)

While the marconi or Bermudian rig was used in Europe on yachts prior to World War I, it did not become popular until 1919 and 1920 on the smaller boats, and shortly after that on the larger ones.

By about 1926 marconi or Bermudian sail plans started to get so high and narrow that the heavier displacement boats found they could no longer tack to leeward with success, and that sailing down the wind was a distinctly slow procedure. Bases of fore triangles, and also spinnaker poles, were shortened to an extent where it was almost impossible to set a good-sized spinnaker, and as the width across the foot was shorter than we had been used to looking at, it accentuated the narrowness up at the head.

In the fall of 1927, Sven Salen brought to Long Island Sound his 6-Metre yacht *Maybe,* in which, while racing there, he won the Gold Cup against strong competition. He had the idea that a spinnaker whose leach and luff were identical in length and roping would be a great asset, for in jibing the sail could be just swung over, the leach becoming the luff, and vice versa. The whole maneuver could be carried out without collapsing the sail, thereby saving much valuable time. Such a sail was made, and tried during the Gold Cup series. It was a "Double Spinnaker," and had to be cut with a seam up through the middle to get the desired effect of having both sides identical. Up to this time, spinnakers were almost universally made with the cloths set parallel to the leach, and only the luff was roped. It was found in making this new double spinnaker that: (1) It could be made much wider or fuller at the head than a normal spinnaker; (2) That it had to be made wider on the foot than usual in order to be set out and around the jib stay.

Ordinary spinnakers that had been previously made as long on the foot had been condemned because they would not lift, and just hung, a mere mass of canvas half over the fore deck and the rest trailing down in the water.

The interesting part about this double spinnaker was the manner in which the fuller head lifted the sail when set, and one did not find the width of foot excessive. Mr. Salen even used this sail reaching to good advantage.

Yachtsmen in America quickly realized the value of this type of sail, and also the fact that a method of circumventing existing rules had been found whereby one could really have a spinnaker to pull one's boat down wind. These double spinnakers were made larger and larger, and finally the 6-Metres were setting them having an area of over twice their regular rated sail area. These were called "Circus Tents" (a good name, too, for you could wrap the boat up in one). Cruising boats with 750 square feet in mainsail and headsails were carrying over 3000 square feet in the spinnaker when they were racing. These were great days, and it certainly was a lot of fun to sit at a tiller and literally feel the old boat get up and go places.

In 1931 the United States 6-Metre team, composed of *Bob-Cat, Jill, Lucie* and *Nancy,* visited the Solent. Apart from good work to windward, what they did off the wind must have been, to say the least, discouraging to the British team. The American team used their Circus Tents to maximum advantage running and reaching, and the series resulted in a most decisive American victory. The following year the first move to limit the size of spinnakers was started by the Y.R.A. in London, and this has been followed by other associations and districts in America since. Few districts have escaped these limiting rules.

It was always easy to measure the height of the spinnaker block above the deck and the length of the spinnaker pole, and allow anything to be set, but scheming out a rule and actually measuring the size of these elastic, easy-stretching, light balloon sails is fraught with difficulty. Just try to measure one of these sails with 15,000 square feet in it!

A rule for limitation of spinnaker size which will satisfy a man whose schooner has a short foremast and relatively long spinnaker pole will draw much fire from a man whose cutter or yawl has a tall spar with the spinnaker set to the top and a relatively short spinnaker pole. "Too narrow a sail," they cry, "it oscillates, and would be steadier at sea if a little fuller aloft." With this I agree, but if the rule is made to suit the latter sail plan, the man with the former plan says: "Why give these fellows with a tall, narrow rig privileges which I cannot take advantage of—the old rule was all right. I can't use any larger and fuller spinnaker; I can't get any speed running anyhow with my short foremast and small spinnaker!"

And so, there we are, at the present moment, in the midst of several questions. Shall this or that rule be adopted? Or shall there be no limit?

Problems dealing with hull restrictions which seemed far greater than these spinnaker measurement difficulties have been overcome in the past, and I have no doubt that within a few years we shall have another final rule which sailmakers will try to find a way to circumvent!

The question often arises: "Why put round holes in a spinnaker?" One theory is that a boat with a spinnaker set is pushing a cushion of air in front of the spinnaker, and that with holes cut in the sail, the air is allowed to go through it and help displace the cushion. Another idea is that the holes (similar to a hole in a real parachute) stop the oscillating. Other people feel that they can unfortunately tear enough holes in their spinnakers without having them put in by the sailmakers!

All joking aside, however, the art of setting and taking in over-sized spinnakers on boats, before the rules arrived, has developed and left us with many gadgets and methods useful for handling the modern parachute spinnaker which has been developed from the Circus Tent and the first double spinnaker.

The following gadgets and ideas are all open to criticism, but they give what is generally considered to be the best and easiest way of handling "Parachutes."

I have taken the 6-Metre and 8-Metre classes to give a mental picture of the sizes of sails in question, referred to in the following remarks:

(1) *Spinnaker pole and mast fitting.* For 6-Metres and other small craft, have the spinnaker pole made double-ended, with a spring snap fitting on each end (stock fittings are now available for this purpose). Place two pad eyes on the mast, at different heights, for different wind conditions and different points of sailing. On 8-Metres and larger craft, replace the pad eyes with a track with holes in it. Equip track with a slide, which latter is itself equipped with a spring snap. The pole may thus instantly be set at any desired height on track. On fairly large yachts, say 12-Metres and up, the spring snaps are not strong enough for the job, so an old-fashioned type bronze cup is used on the slide, which goes up and down the fore side of the mast.

All spinnaker poles should have a strop for a topping lift. On small craft, it should be in the center of the pole, and on larger boats, just outside the balancing point. One of the benefits gained from a parachute spinnaker when properly set is that it lifts the head of the boat. When I say "properly set" I mean that the sail should balloon out ahead of the boat. The outboard end of the pole should be kept up in the air, and the lift should take the weight of the pole. If the weight of the pole comes on the sail and pulls down the stay, the sail cannot balloon out ahead effectively. Conversely, when the spinnaker collapses (which these parachutes have an annoying habit of doing), if there is no lift on the pole, the pole naturally drops down on deck or in the water, so that when the sail fills again you are very apt to tear the sail and probably break the pole as well.

A forward guy is advised for small boats but is not essential. It is when your spinnaker collapses that you will find a forward guy very handy to keep the pole in place.

(2) *Guys, Sheets, etc., Jibing.* On 6-Metres and 8-Metres, where spinnaker tack outhauls are not used and the sail is jibed without being taken in (the same applies to 12-Metres in light weather when doing likewise) one should have two after guys, or sheets, each with a swivel snap shackle spliced in one end. Each guy should be just a bit longer than the over-all length of the boat (this may seem long, but you'll see why shortly). Both should be hooked on to the sail, one on each tack. The sail is hoisted "two blocks" on the halliard (either in stops or loose). One end of the pole is snapped into the swivel of the snap shackle on one of the guys, the other end of the pole is snapped into the socket, or eye, on the slide on the mast track, and the sail is ready for setting.

The spinnaker boom topping lift keeps the pole almost parallel to the water line, the outer end being tipped down a bit to keep the pole from skyrocketing. A forward guy, while not essential, is advised even on small boats. It is a nuisance to have this extra rope in the way when setting the sail, so it can be snapped on afterwards if wanted. The forward guy, like the after guys, has a snap shackle on one end, so that it can be snapped right on to the after guy and allowed to run along it until it jams up against the swivel on the after guy. It can be led forward, then aft to the cockpit for easy tending.

A *swivel* snap shackle is essential on the after guys, or sheets, to allow turns to run out of the spinnaker stay if there happen to be some in it.

How high up on the track on the mast the pole should be set depends upon the strength of the breeze, and whether the boat is on a dead run, quartering run, or reach. As previously mentioned, have the after guys (or sheets) just longer than the length of the boat. It is usually best to have the sheet lead well aft.

Now for a jibe:

Slack up forward guy, let snap shackle run in, then unhook it. Unhook pole from mast fitting and let pole come inboard. Unhook pole from snap shackle on after guy. Hook opposite end of pole into snap shackle on what was previously the sheet, and now becomes an after guy—the original after guy now becomes the sheet. Hook pole into mast slide.

Actually, what we have done is to reverse the ends of the pole—the former outboard end is now on the mast, and the end which was on the mast is now outboard—and have changed a sheet into a guy, and vice versa. While the spinnaker is being jibed, the main boom is coming over slowly, thereby keeping wind in the spinnaker. Hook on your forward guy, set it up, and all should be serene.

(3) Where the gear is too heavy to jibe the spinnaker all standing, two spinnaker poles are suggested, one on each side of the deck, each with its own after guy, forward guy, and outhaul all rove off. Then in case of a jibe, get the spinnaker off, take in the pole, get the other pole out and hold it in position with the topping lift and both guys pulled taut, and set the spinnaker again. (It is easier to have two spinnakers, and then one of them can always be stopped up!)

(4) *Stopping up and breaking out parachutes.* The following can apply to small as well as large sails. The breaking line can be dispensed with in the smaller ones. The first rule for stopping up spinnakers, or any light sails, is *never roll them up*. Flake, or gather the canvas up, then tie the stops around.

On old-fashioned spinnakers, the leach was laid along the luff and the sail gathered up and stopped, leaving all the foot and the lower part of the sail in a bulky bundle. Forget this procedure. Place the two tacks together, flake or gather up the bulky part of the sail, and the "bundle" will be well up aloft. Before stopping, prepare a "breaking line"—a small-diameter rope, with a bowline in one end. Lay the breaking line along the flaked sail, bowline at top. Tie the top stop through the bowline. Half-hitch other stops to breaking line until you reach the point where the middle of the foot of the sail is. Below this last stop tied to breaking line, tie stops around sail in the old-fashioned way.

Having sail stopped in this way, masthead it on the halliard, taking care to follow it up and minimize the turns in it. If one man follows the sail up as hoisted, it is possible to get it up to the mast head without any turns in it. It is advisable not to square the pole too much at first. Hold it in place with the guys and lift at an angle of about 45 degrees forward, its outer end tipped down a bit and about 5 feet (in 6-Metres) above the water line. The outhaul is then hooked onto the tack, *two* sheets hooked onto the clew and sail pulled out on the outhaul to end of spinnaker pole. As to the two sheets, one should come in directly on the weather side of the boat, and the other should be passed out around the headstay, or topmast stay, and aboard again, just abaft the main shrouds. When this is done the sail is ready to break out.

Instead of first pulling on the windward sheet to break the sail out, pull on the breaking line. Here is where the bowline in the top of the breaking line comes in, for if some of the lower stops slip up the breaking line, by pulling quickly the bowline breaks them as it comes down. Theoretically, all upper stops should break together. Once the head of the sail is broken out the windward sheet can be pulled, and you will find that as the lower part of the sail drops, the wind catches it before it drops down into the water, or gets under the bowsprit. Care should be taken to trim in the other sheet, which has already been passed around the stays, immediately the wind starts to fill out the sail. Discretion has to be used in hauling in the sheet at this particular moment—it depends upon the strength of the wind as to how quickly it can be accomplished. The windward sheet is then passed out around the stay and the sail is trimmed by using both sheets.

(5) *Taking in large-sized spinnakers.* The most important thing is to have a man at the wheel who knows how to steer a boat practically running by the lee.

(1) If the pole is squared aft and it is not blowing too hard, ease pole forward to an angle of about 45 degrees. If breeze is too fresh, leave it where it is.

(2) Get your crew gathered in the lee rigging.

(3) Run the boat off until the spinnaker is practically becalmed behind the mainsail.

(4) Ease off on the spinnaker outhaul, letting the sail swing right

around outside everything to leeward (let the outhaul go if necessary; it will unreave itself and as it's made fast to the sail at the tack, you can drag it in later).

At the same time the crew gathers in the foot. Then, *and only then,* lower away on the halliard. The mate should not sing out to the helmsman to luff back onto his course until the head of the sail is on deck. If the boat is luffed before this, you will either lose the spinnaker or drag the crew overboard.

These large sails are not hard to handle if these suggestions are carried out. It is a great mistake, I feel (and in a breeze of wind an impossibility), to take these sails in to windward of the headstay. You lose so little running the boat off for a moment, and you can get the sail in without tearing it.

A Single-Handed Atlantic Passage

MARIN-MARIE

AFTER reading such accounts as those of *Dorade's* crossing of the North Atlantic, or Carl Weagant's voyage to the Mediterranean in *Carlsark,* one cannot help feeling how difficult it is to measure up to the standard set by these yarns. Anyhow, among American yachtsmen, interest in "short boats for long cruises" is so keen that I feel I have some excuse for writing, in my turn, of my lonely crossing of the Atlantic from east to west.

Some people find it surprising that anyone should undertake such extensive cruises, and especially that anyone should do so alone. The question most often put to Captain Slocum was: "Why all alone? Not even a dog or a cat?" Scores of times I was asked the same question. Concerning the objection to animals on board, Captain Slocum has given his reasons in his book, "Sailing Alone Around the World." Every time he took an animal with him some disaster occurred; the goat ate his hat and charts, the cat took to devouring the food, a rat bit him, and the centipede ate the oakum in the seams, and he lived in fear of hydrophobia attacking any dog on board. As for me, the truth is that I was far too much concerned about my water supply to think of sharing it with an animal friend ·who would require just as much of it as a human being.

Sailing boats interest me and I not only like to look at them, but to sail and run them myself. It pleases me immensely to put to the test, in a serious manner, certain conceptions of which I am theoretically sure, and to experiment with others of which I am less sure. It is a question of forming an opinion of my own instead of being satisfied with a ready-made one, or of speculating on hastily formed conclusions.

As for taking but one other man with me on a long trip, that was unthinkable. Two is *not* a good number. Few people realize the responsibility of the man who is sleeping down below, while the one on deck is on watch alone. I assure you that it is impossible to sleep peacefully for a single minute. Again, two persons together take greater risks than will one man; they wait longer before changing jibs or taking in reefs. And, in rough weather, should one of them fall overboard, the other is faced with the terrible responsibility of attempting a rescue. It would be no joke to arrive on the other side of the Atlantic, having left one's comrade "in the cider," as we say, or having let him die on board for want of proper doctoring. I should prefer not to arrive at all.

Winibelle II was my boat, a 36-foot double-ender, moderately broad and rather deep, so as to allow enough headroom below while permitting of a flush deck. Except for her stern, she was not like the Norwegian type, the "Colin Archers" being rather beamy craft of moderate draft. She was built in Boulogne in a most substantial manner, was copper fastened and sheathed, with lead keel. Inside, she was nicely fitted—all lacquer and mahogany, even in the fo'c's'le. Originally meant for coastal cruising and an occasional Channel race, she was first fitted with a marconi rig. When first completed, I thought that with very small alterations about her deck, and with a different rig, she would be of the type and build to stand an ocean cruise, not to mention a 'round the world cruise.

The rig was of a very old style, that of French pilot boats. The mainsail loose-footed; no boom at all. Then there were two forestaysails which could be hoisted alternately, a small one and a larger one. I shall be asked why I did not choose, rather, to retain the Bermudian rig. For three very simple reasons. In the first place, I have not had enough experience with the rig, and I am not bent on gaining that particular experience in the middle of the Atlantic. Second, the rigging has to be kept taut if you want it to stand up, and well balanced. On a new boat especially, it is necessary to watch the stretching of wire very closely, which one man cannot do carefully enough at sea. Third, if I have had trouble with booms, I have had none with gaffs, which make it possible to let down the mainsail in five seconds without leaving the helm.

I have come to hate booms and patent reefing. In a calm they are noisy and impose severe chafe and strain on everything aloft. In a gale they are beastly; until you have boom and sail lashed securely on deck, and a trysail rigged, you never feel safe. Then I had to run 3,000 miles in the trades, remember, and alone, with constant danger of jibing. So that wouldn't do. I dropped the boom completely and found that, after all, with a loose-footed mainsail, a boat would haul as close to the wind as any other, surprising as this may seem. And how reliable! Of course, I kept a spare gaff and a spare mainsail. The rig was completed by control of the helm from the interior of the boat and by a special device for steering automatically dead before the wind.

As for the inner accommodations, aft was a separate engine room, with a 10 h.p. full diesel (side propeller) that gave electric light at the same time, and always worked like a clock, never smoking, never failing to start at the first pull of the handle. Down the hatchway was the galley and my berth, well aft—that always was my plan. Then the cabin, and forward, in the fo'c's'le, the sails were kept. Water in separate ballasts, hand pumps, etc.

The wind blew steadily from the SW, chasing gray clouds across the sky and bringing rain. I rang up Paris to get the time from the Observatory to regulate my chronometer. This was quite an affair, and I was victimized by the telephone operators, who, every few seconds, would urge me to speak. They could not be made to understand that I was listening to the automatic machine at the Observatory. Finally, convinced that they dealt with a maniac, they refused to give me the connection!

At seven in the morning of May 7th, a tendency of the wind to shift to the west brought me to an abrupt decision to set out. I put in a reef and rigged a small jib. The weather was cold and wet, and I was in a beastly mood owning to my having caught a chill some days previously. In five minutes I bade farewell to my friend Bob, who had spent the last few nights on board. He was still anxious to go with me and only with great difficulty restrained himself from insisting on doing so. In five minutes he packed off, and stood there on the quay, all alone, watching *Winibelle* sail slowly down the river. It pained me to leave him there, silent and reproachful. . . .

It was exactly a week after leaving Brest that I found myself due east of Finisterre. In another week, at the end of the day, I sighted Porto Santo, thirty miles north of Madeira. I preferred to land there rather than on the island of Madeira itself, which is guarded on the east by the islands of Desertas and Bugio which are not lighted. I lay close-reefed and hove to during the night, between Porto Santo and Madeira, waiting for daylight to enter Funchal, where I came to anchor in the little harbor of Pontinha, among a crowd of small boats and launches. It was early in the morning of May 25th. I had decided to call at Madeira, as I had noticed that the rudder head suffered from too much strain when the tiller was lashed, in spite of the fact that the rake was as small as possible and the rudder head was very strong. I could have strengthened it with bolts and some hoop iron that I had in reserve, but I preferred to have two iron bands forged for tightening, with two flanges, between which the tiller could be bolted; this is the most reliable device I know of.

I might mention that I had devoted an entire day to trying out the two triangular wings which were meant to enable the boat to sail herself dead before the wind, in the trades. Here I was attacking a problem which, to my knowledge, has never been solved in a really practical manner, and many people had predicted that the idea could not be made to work. In reality, I never expected my device to work in rough weather with a heavy sea following, nor in calm weather either; but I wished to put it to the test. At first I was greatly disappointed; the boat was continually luffing to one side or the other and the braces would not run in the blocks. I remember that, in anger, I struck my fist on the railing so hard that I hurt my hand; but this at once made me laugh again and I set about trying other arrangements, all of them temporary, but which finally gave results. The boat ran before the wind all the latter part of the day and all night at 3.5 knots without my standing at the helm.

At Funchal I had the iron bands made for the rudder. The triangular wings were rigged, I had the boat repainted and revarnished on the outside, and the inside ballast rearranged. I had taken 1,200 kilos on board at St. Malo but this, I had found, was too much, so I threw about half of it overboard in the harbor, to the great delight of divers, for it was good lead, and the customs there would not allow me to put it ashore.

I shoved off at noon of June 9th, and headed directly southwest. Land was out of sight before nightfall. The second day following, I rigged the double staysails, and from then until I sighted the Caribbean the boat steered herself. To my great delight, the automatic steering device worked perfectly, and was as effective in calm weather as in a 50-mile wind and rolling sea. Naturally, I was immensely pleased, for I could spend my time attending to such matters as navigating, keeping the boat shipshape, and cooking; I could lie down and doze, if not sleep. Of course, in mild weather the speed was rather low, so, after a day or two, I rigged a spinnaker, having decided to steer while I kept it set. But, as a matter of fact, even then I found steering unnecessary, for in spite of the pull of canvas to one side, the triangular sails continued to steer the boat quite straight, much to my astonishment.

From now on, there was only one thing which troubled me—how to be sparing of my spinnaker, which was of light, machine-sewn cotton, and to prevent chafe of the running rigging by tallowing it and avoiding friction.

As I drew nearer the Antilles, the weather became more and more murky and stormy. It was at the beginning of the season of squalls. I was making for Fort de France, but when I found that I could make out nothing more than three or four miles away, I decided not to head directly for Martinique, as the east coast of that island is a very awkward one, protected as it is by coral reefs which stretch a long way out. I was obliged to make the east coast of Dominica, a coast that is not lighted, but which is quite safe and steep too. There was enough moonlight and, about four o'clock one morning, I sighted land—a black mass soaring up into the clouds. My impression was that it was like Jan Mayen Island in the Arctic Sea, as I first sighted it.

I sighted Porto Santo dead ahead, in a sort of yellowish mist, which for some hours puzzled me very much. Terrible times those, when, after being alone for weeks, one finally begins to doubt everything— reckonings, calculations, and even one's own equilibrium. But now, I must admit that I gave way to my feelings. Ten days of incessant rolling had simply torn my nerves to shreds. For two nights I had taken the precaution of not sleeping for more than an hour at a time, and finally for no more than half an hour, with the result that I used to dream that I sighted land. When I really did catch sight of it, on putting my head up the hatchway after a short nap, there was a sort of black wall standing up before me. I remember that the first thing I did was to try different ways of convincing myself that I was not dreaming. I cried aloud, "No mistake about it this time, this is the other side!" I walked the deck in a veritable fit of excitement, running the risk of falling overboard. Then I tumbled down the cabin ladder, kissed my children's photos, bounded up on deck again to enjoy the spectacle, laughing and weeping at the same time. Now I can't think of this mad pantomime without smiling. In his book, Slocum often mentions that sudden relief, and the physical breakdown which made him, also, lose control. And remember that he was a man over fifty, who had been in command of big ships on all the seas of the world. I suppose it is the strain of going on for a month without seeing anything, not even a steamer's smoke, and not so much as looking at another human being, that produces this outburst, which I for my part am not in the least ashamed of, but which I am now at a loss to explain clearly.

All this time, the boat was forging straight ahead as if to hit the shore; so I had to pull up and set about hauling down and unrigging the "wings" and hoisting the mainsail so as to make for Martinique. Six hours later, I found myself running under the lee of Mt. Pelee. There, profiting by the sheltered position, I spent several hours putting things in order and preparing to anchor. I started the diesel so as to make the Bay of Fort de France, keeping close to land, and arrived there in the middle of the afternoon, with a quarantine flag flying in the shrouds.

A disastrous hurricane passed over Trinidad two days before my arrival in the Antilles. On July 25th, the day when I decided to set out again, the meteorological service warned me not to get under way; another hurricane was coming up from the southeast. I doubled moorings, but nothing happened. By the 27th, there was a pretty hard wind blowing, and I cast off at 11:00 a.m. under short sail. I had decided to try to pass along the weather side of Dominica and round the Désirade by the east, and this for two reasons: once I had rounded Désirade, I should be able to let the boat steer herself, while I slept, whereas, if I passed by Sombrero or between Guadeloupe and Antigua, I should have had to keep watch for at least three nights. Besides, I should be going directly away from the usual track of tropical cyclones. The officers of the *S.S. Alabama* were not of the same opinion. They said that the sea was too rough for me to go to windward against it, that I could not possibly weather the land, and that I had better sail in the shelter of the islands. And, in fact, when I was halfway up the Dominica Channel, at nightfall, I hesitated; there was a heavy sea running and the current set to leeward. No light-houses on this side and no moon.

Still, I decided to trust the boat and try anyway; she behaved very well, and sailed along wonderfully, forging ahead like a locomotive; this was just the sort of sea that suited her deep and heavy hull, short-masted and rigged to stand up to anything. She liked this better than running dead before the wind, for sure! She traveled so well that by daybreak, the following day, I was abreast of Marie Galante, by nine o'clock I was along by Petite Terre and at noon I had rounded Désirade, not more than a gun shot away. A hundred miles in twenty hours . . . and there I was sailing along at ease, with a little more sheet and the tiller lashed.

Just at this time, a storm was raging over the south of Martinique, the boats in the harbor dragged their moorings, and roofs were flying off the houses in Santa Lucia. If I had gone to leeward of the Lesser Antilles I should have had a rough time.

Here are extracts from the log book:

30 July. No flying fish today. Excellent weather this morning. I tried to do a big stroke and I hoisted the light jib (one I had bought second-hand and rather the worse for wear), just before a squall which I thought favorable. It was blown away in a few seconds; nothing left but the bolt ropes. And I had to haul down the mainsail in a deluge. What a beastly place!

Of course, the skin of my burned fingers was torn away by the salt water as I was maneuvering.

I forgot to mention that I have done no steering since I left Désirade. That's the end of it. I don't steer any more. It's too boring and, with a little thinking, you can always manage without.

8th August. Squalls, squalls by the dozen all around, heavy and vertical like the pillars of an immense building. Strange. Very dangerous. I don't like this climate at all. I prefer the long spells of dull weather we get at home.

10th August. Running along under smaller forestaysail. 8:00 p.m. Obliged to take it in.

11th August. Dirty weather. Got the sea anchor ready.

12th August. It seems I am not to have a single day of decent weather. But that doesn't matter; everything holds properly. I am in no anxiety with my "ridiculously" strong fittings.

15th August. Blowing from the NNE, dead ahead, and no mistake. Barometer rising too high. This gets me thinking, but I can't complain about the actual weather.

16th August. Plenty of flotsam and jetsam. Sounded at noon: 55 meters, sandy bottom. Good. 7:00 p.m. 29 meters, sand. Good. (Really I am rather surprised to find just the soundings I was looking for.)

I believe there is a hurricane coming from far away. Barometer too high, horizon too clear, etc. Cirrus clouds at midday. This confirms my opinion. Cirrus do not clear at sunset. Don't like it. Barometer jumped down 4 points. Horrors!

Started the diesel. Decided to try to get shelter before it starts blowing. Discretion is the better part of valor.

Sighted lights of Atlantic City at 7:45 p.m.

17th August. Lightship *Barnegat* at 1:00 a.m. Heavy traffic there. Lighted searchlight. They all keep away. No fear of anybody coming across my way. Seems to scare them. Saw land at dawn (New Jersey). Violent squall from the W. That's the beginning of it! Took in two reefs. No time to lose now!

Wind shifted to the NE. Very annoying. Too near to the land. Quite calm. What is going to happen now?

From five in the morning until noon, I hugged the coast of New Jersey and Sandy Hook. There were lots of fishing craft, full of anglers, who looked with curiosity at the French ensign at the peak and the quarantine flag at half mast. One of them hailed me and inquired where I came from. When I replied that I had come from the English Channel, I was greeted with shouts of sardonic laughter. I made straight for the fairway and started to sail between the buoys. Huge ocean liners came looming out of the mist, swinging round at fifteen knots and heeling over as they did so. I was afraid of their terrific wash but found it long and not troublesome.

Luckily, I was able to take advantage of the tide, and in less than three hours I had done the whole twenty miles of the channel, passed the Narrows, and arrived in quarantine. I did not dare drop anchor because I felt incapable of heaving in the chain all by myself and getting under way in the middle of the stream amid the traffic. I kept under way between the ships and soon the tug belonging to the medical service came alongside and instructed me to enter a little basin constructed of pilewood, which opened close by. As the entrance was narrow, and the current shot across it, I ran in quickly and then found that the inside was too small to come round head to wind. Fortunately, I was able to wind the boat by letting go the anchor in double quick time and hanging on.

Winibelle swung around neatly by the side of a little steamer. The trip was finished.

It was the 17th of August.

There was a crowd of onlookers in summer clothing. The moment I first walked on the landing stage, a little shaky and still in a sailor's suit and sea boots, I was astonished to find that the people there knew all about my trip; the doctor, the Custom House official and the Immigration Officer all received me smiling amiably. In five minutes the ship's papers had been examined, a cupboard sealed, and the certificate of pratique delivered. An old gentleman came up and spoke to me about Slocum, whom he had known well. I admit I was amazed. Shortly afterwards, the Assistant Commissioner of the French Line arrived in a car; he offered me a launch to tow me up to the company's pier. But I decided to power up there by myself, and at five o'clock I weighed anchor and sailed out again, heading for the entrance of the Hudson River. I began to see Miss Liberty, and then, looming out of the fog, the imposing mass of the city buildings, which I had never seen before.

I was rather excited, as I had been when I sighted the Antilles. In the midst of the indescribable traffic, I rushed forward to get everything shipshape, and then aft again to resume steering. I sang everything that came into my head, and ceremoniously raised my hat to the captains of the tugboats, to Miss Liberty, and to all in sight. *Winibelle* cut right across the sterns of lighters, and acrobatically close to the ends of piers, much to the amusement of the good people. I was mad with delight.

About seven o'clock I saw the funnels of the *Leviathan* soaring up above the roofs and, alongside, the tops of the funnels of the *Ile de France*. A hundred sailors from the *Ile de France* were watching from the end of Pier 57. I came alongside the float quietly, neatly passed a hawser forward, another aft to the onlookers, belayed, not saying a word. . . . All well. . . . Nothing to report. . . . Have a cigarette . . . ?

I had been 100 days on my cruise, 65 of them offshore.

⚓

5

Ranging the Maine Coast

ALFRED F. LOOMIS

Hotspur was in good humor when I rejoined her ten months later. She had been well treated by the Camden yacht yard, having been left in the water until fall and launched again before the season grew warm enough to dry her out. June had been rainy and foggy, but the weather relented the fourth day before I arrived, and I found her freshly painted and as sleek and trim as a two-year-old. Her look of demure self-satisfaction was not born, however, of favors just received but of an honor about to be conferred. At long last my wife was resuming her status as shipmate and permament helmsman.

When the little cutter was first torn from her builder's grasp, P.L. had inducted her into the delights of blue-water sailing. Later they and I had cruised alone among the islands of Greece and established, as we thought, a design which would be reembossed as the years rolled inward. But an enlarged and growing family, plus other matters associated with the dull science of economics, had contrived to blur this pattern. Except for a week here and a few odd days there, P.L., *Hotspur,* and I had not cruised alone for several years.

Now I would be an ingrate and a liar if I wrote anything that seemed to detract from the pleasure derived from sailing with my successive shipmates in *Hotspur's* first summer's cruise in Maine. Each and every one of them contributed in companionship and seamanship more than I deserved. But I must make this observation at the risk of appearing ungrateful and at the same time sentimental. When I cruise either single-handed or with friends, each day's sail is a separate entity—pleasant, strenuous, exciting, or lazily satisfying, but distinct from the days preceding and succeeding it. Each evening when we're anchored and snugged down I am in a state of suspended animation, far from home. But when I cruise with P.L., *Hotspur* is my home and a day is a closely integrated part of life.

P.L., with whom I have discussed the subject, says that it is rankest hypocrisy to attempt to elevate it to a philosophic plane. What I mean, she says, is that when she is along I can sprawl in the cockpit admiring the scenery while she prepares the dinner. I should be the last to deny that she is a good cook. And an excellent helmsman. Frequently when the mosquitoes are bad she also washes the dishes.

The afternoon was well advanced, and our first day's run was a short one to Crockett Cove. But first we made another sentimental journey into Pulpit Harbor to prove that I had not overstated its charm, and to observe the fish hawks diving and nourishing another generation of young ones. Some people will tell you that Pulpit Harbor is so cluttered up with summer residences as to have lost its original savor, but I can only say that it is well worth entering, and, although hard to make out when approaching from the southwest, is secure from all winds when once you get inside.

The Moffats were, of course, the magnet that drew us to Crockett Cove, and the next day, when they invited us to a picnic in a little bay leading out of Hurricane Sound, we almost stopped cruising before we had well begun.

At eleven in the morning we regretfully said good-bye. The wind was northerly as we drifted through Leadbetter Narrows and immediately southerly as we entered Hurricane Sound. But there was remarkably little of it from either direction and we spent three lazy hours covering the seven miles to the bell south of Vinalhaven. The wind failed altogether, and as we had Swans Island in mind as a destination we started the motor and got along. There was rain in the air and a threat of fog, and there seemed no point in lingering in this area of off-lying rocks and ledges.

Sandy Moffat had been telling us of a tragic happening off the Green Island which lies west of Leadbetter Island that wrote finis to one of Maine's dwindling fleet of island steamers. The *Katherine,* of Camden, was bound for Vinalhaven with a party of old people, out for a day's change and recreation. The fog was thick, and on a course between the fairway buoy off Fox Islands Thorofare and the bell off Green Island, the steamer stopped to listen for the bell. Silently and swiftly the tide set her down on Seal Ledge, and as all the weight of the passengers was on the upper deck she immediately lay over and filled. Despairing shrieks of her whistle sounding SOS indicated the *Katherine's* position before her fires were flooded, and brought timely aid. Merle Mills, the Moffat boatman, was among those who heard the distress signal and arrived in his launch from a distance of two miles in time to rescue some of the passengers from the wreck. Other boats came, and none of the passengers died of drowning. But some succumbed to the shock of being plunged from safe security into the icy waters of the bay.

With accidents like these occurring to the proficient mariners of this maritime state, it behooves the stranger in Maine's waters to navigate danger spots with keen awareness. *Hotspur* promptly left astern the ledges south of Vinalhaven and at six-knot speed closed with the beacon on Roaring Bull Ledge south of Isle au Haut. Rain began to fall, and as we progressed I lowered sail (including a brand-new jib which I took below) and put the covers on. We passed close enough to Roaring Bull to see the west-going tide circling around it, and when we had gone another quarter mile the motor stopped.

That was simply lovely. With the companionway slide closed to keep the drizzling rain from the ignition, I explored around and found that the ground connection had succumbed to the cumulative vibration of eight years' use and had broken off. It wasn't hard, although it was long, to cut a new terminal from a piece of sheet copper, but when the circuit was completed the motor still refused to work. P.L. now informed me pleasantly that we were being set back on Roaring Bull and I came to the prompt conclusion that this was no time to pursue my mechanical education. Taking the cover off the mainsail I set that and then, hauling the balloon jib from the bin, I got that on the headstay. The surface of the water showed no faintest suggestion of movement of the air, while the club burgee drooped lifeless from the truck. Yet somewhere in this vertical spread of forty feet there was wind enough—and just enough—to stem the tide and move us away from Roaring Bull.

It was certainly one of those nights. The first thing we gave up was Burnt Coat Harbor on Swans Island. The next, when we saw a bank of fog loitering on the seaward horizon, was thought of any harbor. Our primary endeavor for the ensuing twelve hours was to maintain sea room. Intermittently, if the wind replaced the now drenching rain enough to give us visible speed, we tried to make easting, but even this modest ambition was fraught with indecision and repining. Each time the fitful southeasterly was succeeded by a patch of calm I thought uneasily of the unseen islands that we skirted. Whenever the wind came in again I felt satisfied with our offing and sailed the starboard tack. The balloon jib had been handed when darkness fell, and since I was unwilling to drench the new working jib until it had been broken in we sailed only under mainsail and staysail. Alternating at the helm at two-hour intervals, swapping a pair of wet mittens as each relieved the other, we dawdled the whole night through.

Daylight found us southeast of Long Island and, in a reluctantly awakening breeze, able to lay a course between it and Great Duck Island for the Western Way. But it was another seven hours before we had covered the remaining fourteen miles to Northeast Harbor, so the wind, which was now easterly, was something short of gale strength. In the Western Way, the rain having ceased, we set the jib again and so had good steerageway to an anchorage in Northeast off the public wharf. Sleep was now claiming our best attention, but a chandler's runner came so promptly and with such friendly desire to please that we ordered stores and asked him to fetch an ignition expert at once. He came, and the motor's ailment, having first been diagnosed as condenser trouble, was eventually located in the contact points. He left and we turned in, not displeased to hear a rising northeasterly breeze sighing in the rigging.

With no acquaintances' yachts in port and the sight of nobody ashore to give us welcome, we thought by eleven of the next morning that Northeast Harbor was best viewed over the taffrail. Then Hugh Matheson's *Azara* blew in from the eastward, and before she was fairly secured to her permanent mooring Hugh Junior was alongside with an invitation to lunch. Although we had just finished breakfast we thought it best to go over and at least talk about the weather, and when we did that we learned that an easterly gale of sufficient strength to bother a boat of exactly *Hotspur's* size was making up outside.

Next day the "gale" moderated, for despite the hospitable blandishments of our friends we were anxious to keep on cruising. The day, which took us to Roque Island in light airs, with an occasional resort to power, was without incident, except that mirage lifted a distant three-masted schooner to such enormous size and prodigious blackness that it was almost frightening. A two-master closer to us and not distorted by mirage, sailed across the apparition like a tug cruising down the length of the *Queen Mary,* yet presently the light shifted and the three-master was seen in her proper proportions, minute on the eastern horizon.

Our Roque Island anchorage was at Lakemans, where Willard Carver keeps a register of cruising yachtsmen and, because of old-time association with Sandy Moffat, Sandy Nielson, the late Duncan Dana, and other early members of the Cruising Club of America, maintains a particularly sharp lookout for the blue and white burgee. We shoved off after Carver had had a drink with me on board, and had called my attention to a seven-foot spot north of Halifax Island. We were bowling along on an easy reach as we approached this spot, and, sitting on the bowsprit to set the jib, I looked down and saw the bottom rising up to meet the top with alarming alacrity. I got out the words, "Luff her up," and then remembered that Carver had said that with our draft we would clear with two feet to spare but that I'd probably have heart failure in the process. By the time P.L. had luffed and filled away again I had recovered from my shock, and the reef was well astern of us. The water was crystal clear.

We did not go up Machias Bay to the town of that name, although I wish now that we had, as it is one of the most historic in the entire State of Maine. The town was founded only a dozen years before the Revolutionary War, but the Pilgrims had had a trading station there in the sixteen twenties and the river at the falls had long been the scene of lumber operations. Shortly after news of the Battle of Lexington had reached this remote and hardy settlement, a British armed schooner, the *Margaretta,* with Captain Moore in command, convoyed two sloops up the river to obtain lumber for the erection of barracks in Boston. The sloops were loaded without molestation, but when Captain Moore ordered down a recently erected liberty pole the patriots began devising ways and means of restricting the captain's activities.

Led by Benjamin Foster, some sixty of the townsmen met outside Machias on the bank of a little stream to discuss an attempt at capturing the vessels. Finding after protracted argument that there was not entire unanimity of purpose, Foster stepped across the stream—subsequently dubbed Foster's Rubicon—and invited those who felt as he did about it to follow his example. All did, some reluctantly, and the die was cast. In church on Sunday the seizure of Captain Moore was attempted, but he took alarm and made good his escape through a window.

The next day a group of the patriots led by Jeremiah O'Brien took possession of one of the lumber sloops, named the *Unity*, and with only sixty rounds of ammunition among twenty muskets dropped down river to engage the *Margaretta*. Captain Moore, believing, perhaps, that there was more bluff than menace in the *Unity*, ordered the sloop to sheer off and, when his order was not obeyed, opened fire. Being armed chiefly with pitchforks and axes in true revolutionary style, the Machias men saw the necessity for direct action and after grappling the schooner carried her by storm.

In the affray, which was the first naval battle of the war, the British captain was mortally wounded (he was only a youngster, poor lad, whose fiancée awaited him ashore) while others of his command and two Americans were also slain. O'Brien, our first naval hero, transferred the *Margaretta's* armament to the *Unity*, which he rechristened *Liberty*, and a month or so later he and Benjamin Foster added to the initial success by capturing two armed British vessels at Bucks Harbor, in Machias Bay.

Two years later, Machias again came into prominence when four British warships under command of a Sir George Collier attempted to subjugate the town. One of the four, the *Hope*, succeeded in breaking through a wooden boom constructed across the river at its junction with the eastern branch, and her men did some damage to barns and dwellings ashore. But the next day the defenders, aided by Passamaquoddy Indians, began a cross fire of musketry from both banks and the *Hope* dropped down river out of control, her men unable to keep the deck. It must have been a galling experience, for twice the British ship grounded and lay under the fire of bullets and arrows until the rising tide floated her clear. She finally regained the company of her sister ships near the head of the bay, and the fleet departed. Measured by the standards of the day, the repulse of the invasion was eminently successful. The more so since Sir George, on reaching safer waters, reported to his superiors that he had accomplished his mission and wiped the rebels out.

The channel to Fort O'Brien Point (if you wish to inspect a well-kept earthworks) is navigable to yachts of any draft and above the point, four feet can be carried to Machias. But the channel is narrow and it wouldn't be a good idea to ground out at the top of the tide, as the normal fall is plenty.

Anxious to get to the eastward while the westerly weather held—for I always entertain the hope that when I get as far as I'm going in one direction the prevailing wind will turn around and blow me back again—we crossed the mouth of Machias Bay and pushed on to Cutler. In so doing we sailed through Foster Channel (strong current), north of the Libby Islands and south of Cross Island. The bold southern face of the last-named island is notorious for its slack air alternating with violent gusts, and I have read somewhere the request that Down East cruisers report on their experiences in the vicinity. So I'll report that a sudden fresh breeze railed us and wafted us as far as the Double Shot Islands, which lie directly southeast of Cross Island, and that after that it took the motor to get us in to Cutler.

Going in to this hospitable and justly celebrated harbor, I elected to pass south of Little River Island and so am in a position to advise others not to do so unless they like to dodge the broken-off stakes of abandoned fish weirs. We went in without trouble at dead low water (the chart showing a minimum of ten feet), at which time the stumps were all visible above the surface. At half tide or better the way looks clear, and if the seas are scending into the harbor I have a feeling that it wouldn't be very pleasant to drop down on one of these derelict and formidable stumps.

The tide being fair, we stopped only long enough for gas and provisions and then shoved on for West Quoddy Head or for points nearer or farther. As to nearer points, there's nothing along this fifteen-mile stretch of coast that looks good to the stranger in need of a port, although I am told that Haycock Harbor, nine miles beyond Cutler, is preeminent if you have time on your hands and the tide is right and you want to get away from it all. Sandy Neilson and Harold Peters drop in whenever they range the coast, and although it does not look navigable on the chart they say that there's a hole in which you lie and that when the tide goes out a barrier wall cuts you off from the sea. The *Cruising Guide* gives specific directions for getting in and highly recommends it as a "remote little eel rut."

Incidentally, it is here in the entrance to the Bay of Fundy that one plays the tides in dead earnest. All along the coast of Maine it is desirable to arrange your day's run to get maximum help from the current, but here it's the difference between a pleasant sail and a hard grind. Our flood tide carried us all the way to West Quoddy Head, which is the easternmost point of the United States, and off the entrance to Quoddy Roads we saw the ebb running out against a southwesterly swell like porpoises after a shark. Quoddy Roads merges northward with Lubec Narrows, and as the Narrows is the only waterway that I have seen described in the Atlantic sections of the *Coast Pilot* as having an eight-knot current at springs it is well to select with care your time of entering.

The cutter's maximum speed under power is six knots. The tides were not at full spring and I suppose the velocity of the south-going current was about six knots. Even though I sometimes make mistakes in subtracting, I could see at a glance that there was no profit to be gained in Lubec Narrows, and so we proceeded up the Bay of Fundy along the outer coast of Campobello Island. The wind had freshened during the afternoon and was now sending us through the water at a rousing pace. But there was a strong set against us, and with the day ending and fog settling over Grand Manan Island it was an open question whether we would make port while we still had visibility.

Off Scott Head, which is about two miles from the upper end of Campobello, the wind diminished and we barely held our own against the tide. So, as is my habit, I said, "To hell with it," and we made Head Harbor under power and sail. Although Head Harbor is beyond the international boundary line and so outside the scope of this book, it is one of the most secure and fascinating on the coast and should be visited by all who come within striking distance of it. With its narrow entrance masked by Head Harbor Island, it cuts right into the land for more than half a mile and, except for one six-foot spot, is available for any boat of moderate draft. If the six-foot spot sticks in your craw, or if you think it might detain a boat of six-foot or greater draft, remember that down here the range of tides is sixteen to twenty feet and that a good thing is worth waiting for.

We entered south of Head Harbor Island, cutting the motor as the tide let go of us, and for a few optimistic minutes I entertained the idea of beating up the pencil-shaped haven. On the chart it looked a couple of hundred yards wide, but there were two things the chart did not show. The first was a fish weir, extending more than half-way across the entrance, and the second was a boat coming out. I suppose there was more than fifty feet in the clear in which to work our cutter, but it looked like less; so while P.L. held her dead to windward I started the engine and lowered sail.

Besides its remoteness and complete seclusion, Head Harbor has another attraction—a double series of mooring buoys of the size of telephone poles which are available, in summer at least, to visitors. The water of the harbor has the peculiarity of not freezing even in the severest winters. Because of the harbor's freedom from ice (an immunity which is attributable to the terrific churning effect of the Fundy tides) it is much used for the winter storage of tugs and fishing boats. Hence the buoys, most of them vacant in summer.

Out of the fog of the next morning (Grand Manan had evidently sent its consignment over to us) came Leonard Dyer, a Cruising Club man who lives on Head Harbor Island and who had watched our approach the evening before. There is a red buoy off the southwest end of his island which is not shown on the American chart and which is intended to be left to starboard by those who enter as we did, from southward. Dyer told us that the buoy's location is confusing to those who go in the northern way, and that many strangers, suddenly catching sight of it, sail over the rock it guards against in order to leave the buoy to starboard. He added, "Boats from Boston usually do this," but there was no inflection in his voice to indicate whether he spoke in praise or in blame.

The fog hung on, and for a while at lunch we were all excited over the prospect of *Hotspur's* lying over a night so that our hosts could take us to the opening night of the Campobello Yacht Club at Welchpool, two or three miles away on the inner shore of Campobello. It would be a chance to have a good time and meet interesting people, and the delay of a day might give the weather opportunity to make up its mind. But this scheme fell through when inquiries were made by telephone and it was learned that the party had taken place the night before.

At about two-thirty the fog scaled off, and so with hearty thanks and cordial good-byes we shoved off, rounded East Quoddy Head (from which the island takes its present disputed name), and first with sail and then alternately with power and sail we made our way to Little Letite Passage, one of the two eastern entrances to Passamaquoddy Bay. The afternoon was deliciously warm, and the going was delightful whether or not there was air to fill our sails. We passed inside a small archipelago extending from Tinker Island to Adam Island, but before getting to Little Letite I had a mental aberration that gave me considerable pain. I wanted to go through a short cut which is marked by a spindle a couple of hundred yards off Parker Island, and although the way looks clear on the chart and despite the fact that our boat was on an easily identifiable range, I couldn't make the spindle fit into the picture of surrounding rocks and ledges. At the last minute I got cold feet, turned around, and entered Little Letite by a more open channel.

A nice breeze helped us through the passage, and when we had cleared the headlands which identify it we found ourselves in Passamaquoddy Bay. Its broad expanse and its hilly surrounding shore tempt many to liken it to the Penobscot. I merely give this information for what it is worth, for I saw nothing that would hold a candle to the Penobscot. I must admit, however, that we had no time to go up the famous Magaguadavic (Mackadavy) River and, instead, headed with the last of the afternoon breeze for the New Brunswick anchorage and town of St. Andrews.

St. Andrews Harbor is formed by the low and sandy Navy Island, it in turn having been evolved by the meeting of the tides dropping down from the St. Croix River on the west and Passamaquoddy Bay to the north and east. There are lighthouses marking the ends of bars jutting southeast from the harbor, and these somehow produced the illusion that we had been transported to the Chesapeake. But the abrupt little hills of the mainland were typically Down East, and presently, as a fair wind wafted us slowly past Navy Island, we had other evidence that our locale was unchanged. The tide turned against us, gathered strength, began to suck the buoys under, and for half an hour waged a determined battle with the dying breeze. *Hotspur* was the pawn in this noiseless and almost motionless contest, and we gave her every aid we could short of starting the motor or setting the spinnaker. Presently the wind won out, pushing us past the C.P.R. pier, where the current is the strongest, and we drifted on to an anchorage near the center of the harbor. It was a peaceful and satisfactory conclusion to our outward passage.

⚓

6

The Care of Sails

ERNEST RATSEY

ALMOST everyone these days is familiar with the forceful instructions given the purchaser of a new motor car—instructions regarding the extreme care which should be taken in driving a new car slowly and carefully during the "breaking in" period, in order that the car may deliver peak performance over a long period of years. Abuse during the "breaking in" period may easily result in poor performance and quick deterioration of a fine mechanism.

Exactly the same thing is true of a new sail. The set and effectiveness may be utterly spoiled the first time you set it. Carelessness and inattention during the early life of the sail may easily result in a shortened life.

In an effort to get long, effective service from our sails, let us start with a brand-new sail—a jib-headed mainsail, for instance—and treat it in accordance with the knowledge gained over a long period of years.

Choose a fine, sunny day when the breeze is light. Never set a new sail if it is blowing briskly, or on a foggy, rainy, damp day. Never reef a new sail unless it is a matter of life or death.

Hoist away on the luff until it is fairly taut—no harder. Pull the sail out on the boom a little harder than "hand taut." By this, we mean pull it out until the small wrinkles along the foot just disappear. The bolt rope, which stretches more than the canvas, is sewn on the sail fairly taut, causing the wrinkles. When pulled out so that the wrinkles disappear, the sail will then be of exactly the same dimensions as it was when the canvas was measured, cut and sewn together—the shape and size determined by the sailmaker—the "made length."

Pulling out a sail "hand taut" is the popular rule; we venture to say that this is not quite hard enough, and prefer the method whereby it is pulled out until the wrinkles along the roping disappear. Few coils of rope stretch the same amount. Different sizes of rope stretch differently. A 5-pound pull may be enough on one rope to make the wrinkles disappear. It may take a 25-pound pull on another rope. Wherefore, we prefer to be guided by the wrinkles, rather than the questionable "hand taut" method for different sizes and characters of rope.

The leech of your sail is seldom roped. It is usually cut with a convex edge, or "roach," as it is popularly called. Batten pockets, for wooden battens, are placed at intervals along this edge. Never hoist a sail without inserting the battens in their proper pockets. For if they are not used, the "roach" will not hold itself flat, and the weight of the boom will stretch the sail in a straight line from headboard to clew. But when the battens are in place, the whole roached area of the leech will take its share of the strain, and the sail will stretch evenly and naturally throughout its entire area.

Have your battens an inch or so shorter than the pockets in which they fit. If they are too long they will wear a hole in the inboard end of the batten pocket, or through the body of the sail itself.

The unroped leech of a sail will stretch practically all it is ever going to stretch the first time the sail is used. That is why, among other reasons, we recommend that luff and foot should be pulled to the "made length," to help stretch the whole sail evenly. If the luff is not set up to the "made length," the after end of the boom is apt to droop and swing too low, putting undue weight on the leech.

When hoisting sail, take the weight of the boom on the topping lift until the sail is hoisted all the way up—or, in a small boat, have someone hold the boom up. It is not fair treatment to make the sail take the weight of the boom until the halliard is set up and the luff taut.

With the sail properly set, get under way and cruise around aimlessly for an hour or two. Do not let a new sail shake, with the boat head to wind at her mooring, for any longer than necessary. It will get more than enough shaking during its natural life. Allowing the sail to shake and slat means that the minute threads and fibres of cotton are chafing against one another—the natural life of the canvas is being shortened.

While sailing around with your new sail, you will note that the leech area stretches first. Therefore, do not sail hard on the wind with the sheet trimmed in flat. Watch your new sail carefully. Soon you will see that the luff rope and foot rope are beginning to stretch—there are sags and wrinkles along both luff and foot of sail. As this stretching takes place, haul up on your halliard, and haul out on your clew outhaul, an inch or so at a time—just enough to take up the sags and wrinkles, and no more. Don't "winch up" the sail bar taut—don't pull out the foot of the sail bar taut. The canvas is stretching slowly and naturally—due to mild wind pressure. Stretch luff and foot of sail slowly and naturally— don't force things, or your sail may be utterly ruined so far as proper set is concerned.

There is no set rule as to the number of hours of sailing needed to break in a sail properly. Sails of light-weight canvas, as a rule, break in quicker than those of heavy material. And sails are temperamental. We hazard a guess that a sail should have at least four hours of careful sailing and stretching in clear, balmy weather and light air before being used at all in a strong breeze. Again, it may take much longer. Most assuredly, it is safer to overdo the breaking-in period, rather than underdo it.

If any sail, such as a jib, has a wire luff rope, forget all that has been said about gradually stretching the luff—set it up as taut as you please. Your sailmaker has put a wire in that luff because, in designing that sail, he does not wish that particular edge of the sail to stretch at all.

Most of the foregoing advice on marconi sails applies equally well to gaff-headed sails, the recommendations regarding the treatment of the foot of the sail being applicable to the head of a gaff sail.

As almost all sailors know, the most effective form of a sail is one which, generally speaking, is curved in a similar manner to a bird's wing—the forward section of the sail has a distinct curve, or flow, which gradually flattens towards the middle of the sail, ending in a practically flat surface in the region of the leech, or after edge. It is not all unlike the curve so carefully constructed in the wing of a modern airplane. The wing of an airplane, mounted "on end" in a small boat, has, in fact, been made to propel the boat to windward.

In a marconi, or jib-headed, sail, the draft, or flow, is kept in the forward part of the sail by setting up on the halliard so that there is more strain on the luff rope than there is on the foot rope.

In a gaff-headed sail the draft, or flow, is obtained by pulling out the head of the sail along the gaff until some strain comes on the canvas itself, evidenced by the appearance of small wrinkles or folds running parallel to the head rope; then the sail is set, head to wind, luff rope properly tautened, and gaff "peaked up" until small wrinkles or folds appear running from the peak, or gaff end, to the tack of the sail.

After a period of time, depending on weather conditions during which the sail is used and on the amount of stretch allowed by the sailmaker, the head of the sail may stretch beyond the available length of the gaff. Then it should either be re-roped or have a piece cut off it, whichever the sailmaker may decide is best for the sail. The same treatment is called for on the foot of the sail should it stretch beyond the limit established by the boom end.

If the head of the sail is too long for the gaff, the draft or flow in the sail will be too far aft—near the middle section of the sail—a defect generally considered fatal for speed to windward. No amount of "peaking-up" the gaff will help matters.

If the foot of either a gaff or marconi is too long for the boom, the sail cannot be flattened out as much as we might wish, but it will not cause the draft to come too far back in the sail.

If any sail, properly cut by a competent sail-maker, is designed and built to set on straight spars, common sense tells us that it will not set well if the spars bend or buckle when under way. Spars which may be quite straight when the boat is swinging idly at her mooring may take on some unexpected and remarkable curves when she is being sailed in a good breeze.

Anyone at the helm of a boat is in a poor position to determine whether or not his spars are buckling, let alone tell how much or in just what direction.

To find out if your boom is buckling, go forward while someone else is sailing the boat to windward in a smart breeze, squint along the boom with your eye close to the gooseneck, much as you would sight a rifle. You will probably be amazed to see that the boom which you thought quite straight has taken on quite a curve. The remedy lies in rearranging main sheet leads, or bridles, to bring the strain in the right place to cure the buckling. If this treatment is ineffective, probably your boom is too small in diameter, or just too limber—you need a new, stiffer boom.

Now go forward, lie down on deck, and squint along the mast. Curves and buckles? Probably. Careful adjustment of headstays, shrouds and backstays is necessary to cure the "bends" in a mast. Occasionally, a limber mast will have to be replaced by a stiffer one. Or perhaps rigging and spreaders are not placed properly on the mast. In case of extreme difficulty with either mast or boom, call in an expert—but at all costs, keep your spars straight if you expect your sails to set properly.

Spars are usually tapered. Well-made spars are usually tapered on the side opposite to that on which the sail is attached—the forward side of the mast, and under side of the boom. Therefore, when "sighting" spars, see that the after, or "track" side of the mast is in a straight line, not the fore side; the top of the boom should be straight, not the under-side.

The lead of the main halliard as it leaves the mast on the after side should be vertical. This is accomplished by having a halliard sheave of larger diameter than the masthead, and putting the pin of the sheave just abaft the center line of the mast. This results in the sheave protruding from the after end of the spar. If a smaller sheave is used, the halliard is sure to pull the headboard of the sail against the track. This will soon cause the seizing of the topmost mainsail slide to chafe off, and the track will chafe through the top part of the luff rope before long.

If a permanent backstay is fitted, a wooden or metal "crane" device should be attached to the after side of the mast above the sheave, and the stay set over this crane in order that it may not interfere with the head of the sail.

When sighting along the foot of the sail, be sure that the gooseneck fitting, and the clew outhaul fitting, keep tack cringle and clew cringle in a straight line with the foot of the sail. Frequently they do not. Sometimes the gooseneck fitting is too high, rarely too low. In either case, the sail cannot set properly. The same is true of the clew. For instance, most sails, attached to a boom by the use of track and slides, or by means of a jackstay or lacing, will set for almost their entire length in a straight line an inch or so above the boom. Then suddenly, at either or both ends, there is an abrupt angle where the corners of the sail are attached. No sail should be expected to set properly under these circumstances. See that the fittings at both ends of the boom are such that clew and tack cringles are held in a straight line with the foot of the sail. Otherwise, wrinkles will develop and will become permanent if allowed to remain for any length of time.

It is common practice to lace the head of the sail to the gaff either with individual stops, or a lacing half-hitched at every grommet. Some small craft use a track on the gaff with slides on the head of the sail. For any sail in excess of around 200 square feet area this method is not recommended, for owing to the high angle of the gaff the head of the sail stretches, and the slides allow the slack to slide down the gaff and settle near the throat. Bending the sail to the gaff with lacing, properly hitched, tends to hold the sail to the gaff at each grommet, instead of allowing the slack to settle down near the throat, as when track and slides are used.

With loose-footed headsails little breaking-in is necessary. Usually, headsails have wire luff ropes, so that they may be set up hard taut the first time out. However, it is advisable to use them first in light or moderate airs, in fine weather, so that the canvas may take its natural shape. Do not torture a new headsail by using it in a hard wind or on a rainy day. Give it a fair chance to become a well-setting, useful sail during its natural life.

Most headsails nowadays are cut with a mitre seam, and the angle at which the sheet pulls is most important. The mitre seam bisects the angle at the clew, and if an imaginary line is drawn from the deck lead through the clew of the sail, it should strike the stay above the mitre seam. This is true for most jibs and staysails. For genoa jibs, of squarer shape, this line should usually strike just below the mitre seam. Owing to the multitude of shapes and sizes of jibs and staysails it is impossible to make any hard and fast rule—experience, and trial and error, are the best rules.

In order to determine the set of your sails, and the proper lead and trim of sheets, it is distinctly advisable to get aboard some other craft while someone else sails your boat around. You will see things which are in no way apparent when you are aboard your own craft. Take a good, long look, from all angles. You can, most certainly, see many things from your own cockpit. Nevertheless, "see your sails as others see them." Your time will be well spent.

It is astonishing how many people make a perfect botch of reefing a sail. Not only is a badly reefed sail inefficient, but the proper set of the sail may be permanently ruined unless reefing is carefully and properly carried out. Here is the proper method:

Lash down tack earring, keeping it in line with track on boom. Pull out on reel earring on leach of sail until sail is hand taut along the row of reef points, and make fast securely to prevent earring working forward when sail is set. Lash down earring securely. Roll bunt of sail carefully and tie in reef points with reef knots around foot rope of sail and not around boom.

If there are no reef points sewn permanently in the sail, reeve a lace line through the grommets, making a continuous spiral around foot rope and bunt of sail. Do not use hitches of any kind at grommets, as the purpose of the lacing is to allow it to render and distribute strain evenly throughout the length of the reefed area.

If sail is not pulled out properly along the line of reef point, an undue strain will be set up on the canvas at each reef point, and the sail become badly stretched. In extreme cases, the sail may be torn at one or more reef points. At any rate, it will set miserably.

When shaking out a reef under way, never let go tack or clew reef earrings until reef points are cast loose, or lacing cut. Start casting off the reef points in the middle of the sail, and work towards the ends. When reef points have all been let go, cast loose tack lashing first, then let go clew lashing and hoist away, being sure topping lift is set up to take the weight of the boom.

When through sailing, do not fail to shake out a reef just as soon as possible, especially if the sail is wet or damp. Leaving a sail reefed unnecessarily will stretch the sail along the row of reef points, and induce mildew and rot. Reefing never helped the set of sail. Avoid reefing whenever possible.

If you are out sailing, and it begins to rain, or a heavy fog sets in, the canvas will begin to shrink before the roping does. Unless halliards and outhauls are slacked up, bit by bit, the sail cannot shrink evenly—the luff and foot are pulled hard flat, while the body of the sail shrinks naturally. Wherefore, as soon as dampness begins to affect the sails, ease off on halliards and outhauls as soon as possible, as much as the circumstances require, or the set of your sails may be permanently ruined. This is why wet weather is so abhorrent to racing yachtsmen—wet weather is so apt to be harmful to their sails, despite the most careful handling. For sometimes you just can't nurse a sail in the middle of a tight race!

With gaff-headed sails, we just have to take a chance on the head of the sail—it is quite out of reach. But a marconi sail is easily adjusted in changing weather. Intelligent handling in wet weather will definitely prolong the set and shape of the sail.

Regardless of the weather—wet or dry—always slack up on your outhauls when sails are lowered in preparation for stowing. Even though the foot rope may be dry at the time, damp weather may set in—the ropes absorb more dampness than the tightly-furled canvas—let them shrink naturally, or they will become bar-taut and be pulled out beyond their elastic limits when they finally dry out—and the set of your sail is seriously and permanently affected.

Never furl up a wet sail. If you do, you are simply inviting mildew and rot. Bundle it up loosely, so the air can get to it.

The worst possible treatment for a damp or wet sail is to furl it up tight and let the sun beat down on it early in the morning. It will dry quickly on the outside, and steam, mildew, and rot on the inside. Wherefore, air out wet sails at the first possible moment. In light weather, hoist up your sails—moderately taut only on luff, head and foot—let the water drain off, so that the dry atmosphere can get at the entire sail at the same time.

When drying sails, remember that the corners, due to the several thicknesses of canvas reinforcing at these points, take much longer to dry out than the body of the sail. The same is usually true of the roping along the edges. Never pull hard on a damp luff rope or footrope—let them stretch out naturally to their previous length—which they will do as they dry, if you will but give them time.

Sail covers are decidedly useful in keeping sails clean and, in many cases, keep out dampness and rain as well. But do not expect any sail cover to keep a sail absolutely dry. Dampness is sure to work inside the best sail cover ever made, and rain is almost certain to work in around the mast. Wherefore, after a rainy spell, remove your sail cover just as soon as the weather clears up—let dry air and sunlight do their job before mildew and rot set in.

Never cover up a sail which is wet, or even damp. The first rays of sunlight will cause steaming, mildew, and rot.

Particularly on small boats, and sometimes on large ones, sails become soaked with salt spray. Even when dry, sails of light material, impregnated with salt, will almost crackle like a piece of paper when handled. Again, when damp or foggy weather sets in, the salt in the sail quickly absorbs the moisture, and the sail is wet. The part of the sail which is wet will shrink—the rest of the sail will not. Result, a poorly-setting sail.

The remedy for salt-incrusted sails is to rinse them out thoroughly in fresh water. Small sails can be washed out in the bathtub. Larger sails can be "hosed down" with a garden hose on the front lawn. Or, you can send your sails to your sailmaker and have him rinse, scrub and clean them—a practice becoming more popular as yachtsmen become ap-preciative of it.

When laying up your sails for the winter, be sure, first, that they are perfectly dry. Then, after removing the battens, fold them carefully, and store them in a clean, dry place.

7

The Wreck of the "Cimba"

RICHARD MAURY

AT SIX in the morning we made sail for Suva, the schooner heaving into light airs with the studied rhythms of a big ship slow-beating up the wind. With Veti Levu in sight, faint ghosts stirred from the southeast, passed, and the southern sky became cross-feathered by cirro-cumulus cloud; the headlands darkened under more compact formations until mountain ranges shone a violent green under walls of black, hanging thunderheads. The scene lightened, the sun glowed weakly, and expecting an east wind we stood in that direction to avoid a beat through Suva Pass. Some rain rattled on the ocean, blew off; the sun dropped, the twilight passed and a long swell rolled in from the south.

In the dark, the east wind began to blow, and we worked into its eye, crawling at perhaps two knots. At ten o'clock, the Suva beacon and harbor range light were raised, but as the wind swung to the northeast we kept tacking east to gain weathering before running the passage. We asked our passengers, young Gordon Griffen and Hugh La Forrest, to go below until we were clear. Shortly later, speed dropped, and it was almost one in the morning before the schooner had run out the required distance. Coming about quietly, we steered for Suva, eight miles off, the wind now spinning from the north. We took ample bearings on Nasalai Light, as well as on the harbor beacons slowly drawing in line, as for the first time on these seas we navigated with the aid of more than one lighthouse revealing an exact position. With Suva five miles off, and the Great Suva Reef two to the north, it began to rain. Thunder sounded on the coast, and the ship was set about by a downpour, a solid wall between shore and sea, a torrential cloudburst cutting off the land wind, the loom of the lights. Suva was abandoned, and we fell south of west for sea room.

Leaving the steering well, Taggart and I took up on the sheets sagging overside, then went to the main rigging to watch for a change. Nothing sounded above the noise of the rain. The sails could not shift the hull, and the engine, repaired no less than nine times since Tahiti, was definitely silenced. While sensing no particular danger, we were aroused by the craft's helplessness and felt of the anchors to make sure they were clear, knowing as we did so that the coast was utterly sheer with the hundred-fathom curve immediately off the reef. When an hour of watch had gone by, Taggart, shouting something above the noise, disappeared below to look at the chart. I left the shrouds and stood on the storehouse top, just forward of the helm: the compass, shining through the wet deadlight at my feet, indicated we still headed for sea.

At approximately two-thirty, I thought I heard a sound above the rain coming from astern. I looked over the sea, my eyes peering to the edge of a circle of visibility perhaps ten feet through. The downpour was deafening, solid. The sound did not return. But as I wiped the rain from my eyes I saw the ghost of foam.

I moved for the anchors and was passing the mainmast when the ship struck on a reef. The deck dropped underfoot, the fore boom drove against my shoulder—and I had fallen through the companion and onto Taggart, sprawled over the chart. There was a loud roar; the cabin heeled, a barrelful of sea foamed over the hatchway, and in one heap, the sleeping passengers, bunkboards, blankets and gear fell to pin us underneath. The sound of coral ripping the underbody followed. There was a deafening concussion, and as the hull skidded over shoal, I found the companion and made the deck, the others behind. But there was no deck left—only a steep incline piling out of the surf. The sails were in the sea, the lee rigging under; the nearest land, two miles off.

We swarmed forward, gripping to the weather rail, up-and-down over our heads. A breaker lifted; there came the cry of "Hold on!" and it pitched against the side, thundered, and drove us through white surf. I reached the fifty-five-pound bower, but young Griffen, who was heavier, took it and threw it perhaps further than any other anchor had been thrown. Griffen fell in the sea. Two of us got him. I had La Forrest climb into the cabin and see that all lights were out. He returned; nothing was on fire, but "She's filling." As a slight wind began to blow, the seventy-five pounder was made ready, its cable hauled out of the wash and laid in bights over the horizontal foremast. Darkness bound sea and ship, the rain increased, the surf sucked, riding up the sides. All halliards were cast off, the sails being hove down the masts on the run, to offset the broadside of wind. Five times an attempt was made to kedge the anchor out to windward. It could not be accomplished in that sea. We managed to parcel the cable of the smaller hook that, cutting a twenty-yard groove through the coral, had yet to part.

A heavy breaker swept the rail, forcing us into the masts. The hull vibrated, the rigging clattered. We had done what we could for the craft and it was time to consider our guests. A dash to leeward might be made in the small tender, too fragile to face the surf. The sea would be quieter under the lee of the reef, where a run for the land could be started. Griffen and La Forrest made ready. We shouted in their ears to send a tugboat—a government tug, not one demanding immediate salvage. Strange how we insisted on a future even while aware that not one of the craft belonging to the five hundred souls drowned by the reef had gotten off. Gordon Griffen, a good young boy, wrung my hand savagely—and he and La Forrest were away.—"Good luck!" someone shouted.

An explosion of surf raked the schooner, drowning her in spray. She yanked at her hook and together with the anchor, gave way into the coral. After some effort, the manila cable, sweated steel-tight, was parceled with additional chafing gear where it bit the chock. Then, in silence, Taggart and I struggled to get sails and booms out of the wash and secured with heavy lashing to handrails, now serving as footwalks. The rain eased and we looked at the black sea, colder than before, and whipped by a drive of wind. There was nothing in sight. The starboard side of the hull lay under water, the port was taking the brunt of the riding sea; however, influenced by the anchor, the hull, which had struck almost broadside, was beginning to point into the waves.

An examination below decks was started on finding a watertight matchbox with which to light the lamp, that with chimney gone, showed the cabin in ruin, the starboard side submerged, the port so high in the air that we stood with water to our knees, not on the flooring, but on the starboard locker shelves. Two tin boxes, one floating some of Taggart's gear, the other, ship's paper, were thrown into a dry locker, holding the chronometer and sextant. Armed with a hurricane lantern, we worked down into the storeroom to find the engine, a thousand rounds of ammunition, and the tanks submerged by a rolling sea, heavy with oil and floating the best part of a hundred charts. As the skin between cabins and planking had not been broached, no leaks could be discovered. We threw off the costly weather coats recently presented by a moving picture company, and bared to the waist, set about with buckets and five-gallon tins to bail ship. More than once we were flung to our knees by the twisting of the hull, but for fifteen, twenty minutes, worked with a will, only to discover in the end that no matter what we did, the water reached sea level. Quite suddenly the cabin made a heavy lunge to leeward, the light went out, and we found ourselves under the slop, hemmed against the bulkheads. The anchor had finally parted!

We were on deck again, laying out the last hook—although we scarcely knew how it was to be done. Because this anchor had been too heavy to cast far enough for a purchase, it had been kept aboard. Now we must attempt to make it hold. While paying it over the bow the wet wind strengthened from the ocean. Climbing down the sloping deck, Taggart and I gripped arms and slowly lowered ourselves into the sea. The wash, waist-high, was powerful, and several minutes passed before we were under the bows, feeling for the anchor. Each man lifting a fluke, we began a slow march out to windward, where the grapple might take a hold. The rain closed in. More than once we were put head-under, while one wave set us swimming. We turned to see how it would hit the craft. She rolled down until her mast tops seemed to strike water, then—as a hard pull was felt on the cable—she disappeared. Throwing a fluke into a pocket of coral, we groped in the sea for the cable, and on holding to it were swept to leeward where finally the hull loomed out of the darkness. We gained it, and swinging aboard, took up on the line, belaying it to the foremast foot with a round turn and two half hitches.

It was almost four o'clock when the smoky lamp below was relit. The grounding had occurred at about two-thirty and since then we had pounded—literally so—over fifty yards of coral. There was no chance of slipping off the lee edge, for at this point the reef was a mile through. And now a new danger was arising: the tide, that had been steadily retreating, cushioned the effect of the rollers less and less, and the hull, picked up by breakers rushing to leeward, was being left to fall solidly onto rock. It seemed that a total breaking-up must shortly take place. A tin of flares was fetched from the ruin in the eyes. A lighted flare might bring aid—but perhaps only the aid of commercial wreckers. Again the insistence on a future!—a gesture I had admired in Carrol, in Warren, in Dombey, and now beheld in George Taggart, as throughout the entire night we worked side by side, sharing the best with the worst. The cabin rocked and echoed with the hollow sound of choking water; the ship's bell tolled, and the yellow lamp sent waves of shadows touching the wet brow, the half-submerged and glorious hair of Miss Landi, and the white spars of our old namesake, sailing upside down on the port bulkhead. Under the dark water were the stoves, the chart table, and floating above them, fragments of books, charts, papers. A paper suitcase dissolved, and I saw for a last time my sentimental poems, washed page by page into the oblivion beyond the hatch. I felt no regret. We cannot hold the same poetry throughout life. There must ever be change, slow or sudden; a change that breaks as one wild wave, or as a slow tide covering over that which is endeared, which, even as we love, we lose.

The tide had turned, but we were still hanging on when the light broke. The headland was seen some two miles away, separated from the wide reef by the breadth of Lauthala Water. The schooner lay on a sharp heel sixty yards from the open sea to the south, and the anchor (we had surged the cable at the mast throughout) still gripped. There was no sign of aid and we grew apprehensive for our passengers.

At six o'clock, the pilot cutter *Mona* drew up on the weather side of the reef, launched a surf boat, and Port Captain Nysmyth boarded us. Ways and means of getting clear were considered. The Captain explained the current that had grounded us during the calm as a freak only occasionally setting into Lauthala Water, rare enough to be omitted by all pilot directions, sudden enough to have been non-existent before the rains came at one o'clock—but strong enough to have set many craft, both local and foreign, onto the reef. During the past fifty years not one of these vessels had gotten off, each one being destroyed, the longest to last being a crack American schooner, a four-master with six-inch planking that had survived the surf eight hours before breaking up. We suggested chopping out the masts, unshipping the iron keel, and working her off the lee, but Captain Nysmyth shook his head, declaring it impractical. After we had declined his offer to board the cutter, being both a true sailor and a gentleman, he set out for Suva to get a tug.

At seven a large white canoe came over the lee. Its owner very kindly took our chronometer and sextant, and told of the night creating a hero. It seemed that after leaving us, Griffen and La Forrest were snagged in coral—rotted areas infested with octopus and water snakes—gashing their feet badly as they waded for Lauthala Harbor. Before long La Forrest severed a tendon, and for a full quarter of a mile, young Gordon carried him on his shoulders, while still towing the dinghy. On reaching open water, the nearly unconscious La Forrest fell into the tender, breaking one of the two-foot oars. With the oar patched, Griffen worked for more than an hour before the tender struck mud, when again, with the other on his back, he waded a full quarter of a mile. On reaching dry land he made La Forrest as comfortable as possible, then ran for almost a mile to the nearest plantation, reporting us shortly before dawn.

With the sea rising, Taggart and I, by using the fore gaff and a wing boom for heavers, tried working a tarpaulin under the stove-in starboard bilges. But the battered side, pounding on shoal, could not be lifted. At nine-thirty, we hove the hawser of government towboat *Number 6* (the most powerful tug in the waters), through the surf and bent it to the mainmast. The tug opened up at full speed, but the hull did not move. We signaled to ease off, and bent the towline through the propeller aperture to obtain a fairer lead. The tug changed direction to break the groove of the keel. Finally the hull moved a foot—a yard—and just then the hawser parted midway. Five times a boat was swamped before a new six-inch manila had been made fast and the tug could pull once more. A silent mob gathered at her bitts watching her strain the tow rope to breaking point as she strove to move us. But it was a vain struggle. The tide had been lost; and in the end the line was cast off. "That's the finish," someone exclaimed.

"We'll try on tomorrow's tide," said Captain Nysmyth. "Get all valuables off, though, I advise." He turned to the owner of the white canoe. "Mr. Turner, will you see these two boys ashore before they pass out? Make them take it easy."

At a big house by the sea, where we could view the craft, we had the first meal since leaving Kandavu—having anticipated a large one upon arriving at Suva—and telephoned time and again to make sure of the aid of *Number 6* on the following day. We went back to work, and aided by the crews of the *Sigawale* and the New Zealander *Arethusa,* our old mooring mates of Suva, strapped eight empty oil drums under the bilges, to buoy the schooner against the morning tide. The men worked with a will, not one believing that there would be any morning. Even if there was, where was the craft that could be mauled by a tow boat over sixty yards of reef without losing her bottom? Already she had outlasted the big American by some hours. Night was closing, the breakers were building. The hull was stripped of sails and tackle, the anchor was set out further, the parted tow rope serving as a new cable. Companions were battened down, the forehatch muzzled with a canvas hood, and then in a wilderness of sunset, the men, laughing good-naturedly, carried us bodily away.

Dawn showed the sea bursting, the hull standing, doggedly opposing it. We were on board in short order. The news reached Suva. *Number 6,* manned by thirty volunteers, skippered by Captain Nysmyth himself, steamed out. The Boys' Grammar School declared a holiday to aid us; the Sea Scouts appeared in the old champion *Heather,* and members of a yacht club, who needed every day's pay, appeared on the scene, joined by the crew of a Polish yawl we had known in Panama. With a little cluster of shipping under our lee and the faithful *Number 6* out to windward, we made ready for a last attempt. The tug started, ten or twelve men heaved on either side of the crippled hull, but, at the end of half an hour, not even a foot had been gained. The tide began to change. During a pause, the oil drums were driven lower into the sea, the masts hove down by a whip leading from the main truck to a kedge. Once more the tug opened up at full, while the crowd swarmed under the bilges. Suddenly we moved, ever so little. There was a sound of crushing coral; a roller came in, driving the men from the sides; but they came back, shouting, heaving. The hull staggered again. Never letting up, the tug strained the hawser bar-taut, the schooner vibrated, moved slowly, pounding toward the surf line, crushing rock every yard of the way. The spars began to righten, both rails came flush with the sea as she lifted, crashed a way for open water. At the lip of the reef, Taggart and I climbed on board, calling for volunteer bailers. Everyone offered; many had to be kept back. The hull dove through the tower of a comber, slumped back, and then with a final pitch lunged onto the sea.

The two of us in the cabin, our heads just clear of the water beneath the carlings, began bailing to keep the craft afloat. The youngsters of the yacht club broached a way into the cabins and, fresh and full of vigor, relieved us of the big oil cans and set to work like Trojans. At either cabin they worked, two men to a tin, slinging the cans on deck, where another two emptied them, threw them back. Gordon Griffen, his legs in bandages, and smiling through tears, was one of them. They began to sing, their work quickened, they gained on the inflow, and the level began to fall. We would not sink, at any rate! Up ahead the *Number 6,* with an occasional blast of her whistle, pulled for Suva with foaming bows.

There were so many to thank! So many! Captain Nysmyth, who had gone so far beyond his official duty in rendering both technical and spiritual aid; Patrick Ewins of the *Sigawale,* almost wholly responsible for the maneuvers of the men in the surf; A. H. Pickmere of the smart *Arethusa,* Superintendent Sabin and Alexander Bentley and their men; P. T. Tucker and his son, whose home had been our recent headquarters, the men of the Polish *Zjawa,* the Grammar School boys, and lastly the men and boys of that remarkable, hard-sailing, salt-water yacht club of Suva.

Some way off the·Public Works Slip the tug let go and we purposely drifted in to ground on the mud. The hull touched bottom, and with a loud exultant cheer, the bailers ceased work, rushed on deck, swam ashore. It was then that the *Cimba* sank, disappearing entirely from sight.

Grapples were fastened to her, rollers were eased under the sides. When the tide dropped, she appeared as though in death, chalk white, buried in mud. Taggart and I stripped her clean, saving some metal equipment, rope, a few clothes, one or two books, which together with saturated charts and a few miscellaneous tins we dried on the shore. When the tide had dropped yet further, a survey of the hull started, a survey which revealed the strength, the integrity, and the honesty of Indian Point ships. True, the false keeling had disappeared; seven strakes had been *rubbed* through by the chafe of coral, and one strake had been pierced. But not a thing more! Not a seam started, not a frame fractured. It was the most incredible fact of the entire voyage!

⚓

8

The "Older Navigation"

HENRY HOWARD

DURING the past thirty-five years much has been written and an enormous amount of work done in perfecting and simplifying the calculations required to determine one's position from observations of the heavenly bodies. These methods have been frequently called the "New Navigation." Is it not well to record and analyze what might be called "Older Navigation," lest it contain certain principles and practices which might be of value to the seamen of today and which are in danger of being forgotten?

Valuable and fascinating information may be obtained by the study of practices and traditions of the natives of the Pacific Islands. The difficulty is that, in many cases, available written records seem to have been made by landlubbers unable either to understand or record intelligently what was common practice among the skillful native navigators when the missionaries first arrived. Perhaps the most skilled were to be found among the Maoris and the Hawaiians.

The knowledge gained by the accumulated experience of the natives would still be of great value to shipwrecked mariners possessed of a life boat but no navigating instruments. This would be especially true of the vast expanses of the Central and Southern Pacific where you may sail day after day without sighting a vessel. The methods used by the native navigators might enable the shipwrecked mariner to locate islands which he would otherwise pass without dreaming of their existence.

Some years ago, I spent the winter in the Hawaiian Islands and discussed the old native methods of navigation with many residents. Traditions carefully handed down describe in much detail long voyages, even as far as Tahiti, and these traditions are supported by the similarity of language, customs and physical characteristics of the people in these widely separated islands.

In the Hawaiian Islands, as well as in many other parts of the world, the natives for generations have studied and learned the apparent movements of the heavenly bodies and used them for maintaining a reasonably true course when out of sight of land. In their long voyages, they used large canoes with large crews. The boats were propelled both by sail and paddle. When they lost sight of one island, they had many ways of determining the direction of other islands if they were not too far away.

The island of Hawaii, with the 14,000-foot volcano, Mauna Loa, could be detected at a distance of 150 miles by the large cloud cap which generally covered the summit and frequently extended much higher. This cloud formation is produced by condensation of the moisture in the Trade Winds. These winds are deflected upward when they reach the slopes of a mountain and, as greater elevation is reached, the temperature drops, producing heavy clouds. The upper part of Mauna Loa is covered with snow and the Trade Wind, rushing up the slope, often carries the clouds formed to a great height.

In the clear atmosphere of the ocean, far removed from the dust and smoke of any continent, such a cloud formation can be seen poking its head above the horizon many miles before the mountain itself becomes visible. The clouds in such case have a peculiar shape which would be instantly recognized by a man skilled in this type of navigation. This principle, of course, applies to any mountainous islands and especially in the Trade Wind belt because of the high moisture content of the warm air.

A somewhat similar method is utilized for locating the low atolls. These usually consist of an extensive shallow lagoon, more or less surrounded by low land frequently not more than five to ten feet above sea level. In many cases, this land is not visible from a canoe from a distance of more than four or five miles, the first thing to be seen being cocoanut trees apparently growing out of the water. However, the presence of such an island can frequently be detected at a distance of 20 to 30 miles by the fact that a small fleecy white cloud is often seen suspended over the atoll during the middle of the day and early afternoon. This cloud is apparently caused by the much higher temperature of the shallow water of the lagoon which causes the air in contact with it to be raised to a higher temperature than the air over the deep water of the surrounding cooler ocean. It also causes more rapid evaporation so that this warmer air also contains an excessive quantity of moisture. This warmer moist air, being lighter, rises continuously to a considerable height where it is cooled by contact with the cooler Trade Winds, the excess moisture forming a small isolated cloud which clearly marks the location of the atoll so long as the sun is high and hot enough to produce this effect.

Another interesting phenomenon, which was much used in navigation between islands 50 to 100 miles apart in the Trade Wind belt, is based on the fact that the Trade Wind, blowing day after day from nearly the same direction, produces large and well formed rollers whose axes are at right angles to the direction of the wind. If, however, you notice that these rollers are no longer moving directly with the wind but at a distinct angle with it, you have good reason for assuming that some obstruction to windward has been the cause. This may be a submerged reef but is more probably an island, and here is where the experience and skill of the navigator comes in; from the amount of the deflection, the character of the waves, he would estimate the direction and distance of the land. In addition, he would study the clouds which might indicate unusual disturbances at a distance which, being the forerunners of an approaching storm, could account for the rollers coming from a different direction. In this latter case, I imagine that, in the beginning of the change, the rollers would come simultaneously from the old and new directions; if the change were due to land, it would result in the deflection of the old rollers with no cross rollers.

The deflection of the large rollers works in this way: Suppose we have two islands, A and B, separated by a channel 50 miles wide—island A being directly north of island B. The Trade Wind and rollers normally come from the east but on the northerly shore of island B the rollers will come from the northeast, while on the southerly shore of island A they will come in from the southeast. The reason for this is the friction of the land and shallow water which slows down the ends of the rollers adjoining the land. This causes them to swing around and come in at an angle towards the coast. The result of this deflection is said to be noticeable for a good many miles to leeward (that is, to the westward of both islands) and enables the native navigator to conclude, first, that land exists not many miles to windward and, second, its approximate direction.

Another way in which this principle was used was in crossing from one island to another which was not in sight but the direction of which was known. Such a crossing could be made in cloudy weather with no sun visible and, of course, with no compass.

The native boatmen in crossing from island B to A (north of B) would find deflection of both wind and wave and were accustomed to use the following rules: At first, after leaving island B, steer a course carrying you diagonally, say, four points to windward of the axis of the rollers but, as you approach mid-channel, gradually change your course so that the wind and sea are just enough forward of the beam to allow for leeway and currents; finally, during the last quarter of the crossing, you can drive the boat with wind abeam or even a little abaft the beam. These changes in the apparent course would result in the boat following a relatively straight line between the two islands. I have talked with a friend who, many years ago, had actually crossed a channel in the Hawaiian Islands in a native canoe with native navigators in exactly this manner and had been greatly surprised at the good landfall they made.

This interesting phenomenon causes a confused sea when you are directly to leeward of a small obstruction with rollers approaching from both sides and crossing each other. I spent an uncomfortable night at anchor close to leeward of Great Inagua, West Indies, the westerly end of the island being about 14 miles wide. With a moderate gale from the east blowing directly offshore, we were tossed about all night long by rollers coming from north and south, in opposite directions, around the island. This disturbance would obviously have been noticeable a substantial distance to leeward—how far I do not know; this is where the experience and skill of the old navigator came in. If you were making good a course at about right angles to the Trades (which would mean wind one or two points forward of the beam), this would, I think, be the best course on which to find land by this method. When you notice a deflection of the axes of the rollers until finally they are coming from more or less abaft the beam, with the wind still from the same direction, you can conclude there is land to windward and would then haul close by the wind. You may be too far to leeward to sight it but, if the island is small, you will soon note a confused sea and then the rollers coming at you from straight ahead, indicating that you are getting past the land. This would be the time to tack. In the old days, with a large crew, the natives would, under those conditions, paddle dead to windward to the land.

Another phenomenon which must not be overlooked is the night land breeze to be found principally in the vicinity of mountainous coastlines. Shortly after sundown, the mountain surface begins to cool much more rapidly than the sea and this cool, heavy air starts flowing down the mountain-sides. This pushes back the Trade Wind so that you may often get a fair wind during the night, close to shore, where the wind will blow directly off the land. In this same place, during the day, the Trade Wind might be blowing parallel with the shore and dead against you. On the leeward coast of a large island, the Trade Wind may be pushed back during the day by the sea breeze which is pulled in by the heated air rising from the hot land.

Still another method of locating small isolated islands in mid-ocean without any navigation instruments other than a compass is by the flight of sea birds. The time chosen is in the birds' breeding season. An ample supply of fish is taken and, after the boat runs so that the crew estimate that she is near the island sought, sea birds are attracted by cleaning fish or throwing fish overboard. Then one bird is selected and scraps thrown to it alone so that it has soon eaten to repletion. Then it seizes the next fish to carry back to its young and makes a bee line for the island where its nest is located. The fishermen note carefully the compass course of its flight and follow it. If the island is not quickly located, this process is repeated and a new bearing obtained. These fishermen are said to pick up the islands without much difficulty. In the South Seas, where the compass was not known, the direction of the bird's flight would be noted as compared with the direction in which the sun bore, that from which the large rollers came and the Trade Wind.

The skilled native navigator, in searching for land, would have all of these methods in mind and would keep going on his course until some sign told him that land was not far away and how he should change his course to reach it.

As a boy, I used to go out with a lobsterman while he hauled his pots off the rocks on the west shore of Narragansett Bay between Narragansett Pier and Point Judith. The trend of the coast was north and south, the prevailing wind southwest, but the big rollers seemed always to come in from a southeasterly direction. My friend never bothered to carry a compass and, in the frequent thick fogs, always steered by the heave of the sea. Later, I frequently used this method of steering in a fog. Let us now consider how the traditional voyages between Hawaii and the South Sea Islands could have been made. The voyage presents no extreme difficulty; all that was necessary was to sail about south in a well-manned and seaworthy canoe and keep going. You could hardly avoid picking up islands where the crew could rest and refit.

It was the return voyage which has seemed difficult because of the isolation of the Hawaiian Islands. However, a study of the large-scale charts covering the water between Hawaii and Tahiti indicates that no great gaps would be met that could not be crossed quite easily and with reasonable safety in returning from Tahiti until Palmyra was reached. The longest jump was from Malden Island to Christmas Island, about 370 miles. This, at the rate of three miles per hour, would mean five days but, with a fair wind and sails, this might have been reduced to two days.

From Palmyra, however, the jump was between 500 and 600 miles but here they were enormously helped by high mountains, Mauna Loa, nearly 14,000 feet, on the Island of Hawaii, and Haleakala, 10,000 feet, on the Island of Maui. As we have shown, the cloud formation above these mountains frequently gives notice to the trained eye of their presence for a distance of 150 miles. Now, in a run of 600 miles, an error in the course of 150 miles either side of the correct course is a substantial one—even rather poor courses sailed by the sun and stars ought to come closer than this—and this largest gap, being the one nearest to Hawaii, might reasonably be expected to be the one with which the old navigators were most familiar. Really, all that was necessary to reach the Hawaiian Islands from Palmyra was to make good a course due north by the sun, North star and nearly constant Trade Wind, and it would have been almost impossible to miss them.

This brings us to the so-called "Sacred Calabash" described by Rear Admiral Hugh Rodman (retired) in "The Proceedings of the United States Naval Institute" for August, 1927. This calabash, he said, can be seen in the Bishop Museum, Honolulu.

According to Admiral Rodman, the "Sacred Calabash" was a cylindrical vessel and, when in use, was filled with water to four holes, all at the same level and spaced equally around the calabash. It was kept vertical by keeping the four holes level with the surface of the water in it. The holes were located at such a distance below the rim of the calabash that, when the observer was in Hawaii, Polaris could just be seen above the rim when looking through the hole on the opposite side. This meant that the observer looked 19° above the horizontal because the altitude of Polaris is the approximate latitude of the observer and 19° is the approximate latitude of the Hawaiian Islands. If the star could be seen above the rim, it indicated you were north of 19° latitude.

According to Admiral Rodman, the method used for returning from the South Sea Islands to Hawaii was to head far to the east of the Hawaiian Islands while going north until observations with the "Sacred Calabash" showed they were in the correct latitude and then head west before the Trade Winds—keeping in the correct latitude by means of nightly observations until the islands were sighted.

Admiral Rodman's story was complicated by the fact that he included in the article a photograph of a vessel which is now in the Museum but which does not correspond at all to his description. First, it is not a calabash or gourd but is of wood, hollowed out; second, it does not have four holes at 90° apart but has ten groups of three holes each; third, the angle between the lowest hole and the rim opposite, if the vessel could be used as described, made 11° 27′ instead of 19°, a difference of 7½° or 450 nautical miles. In other words, it would have located the navigators 450 miles too far south! But the apparatus could not possibly have been used as described because the holes were so small that droplets of water would fill them and prevent any vision through them. They are between ⅟₁₆- and ⅛-inch diameter. I have seen and handled this vessel; fourth, the history of this particular vessel is perfectly well known. The holes were for things to hold the cover in place and it was probably used as a trunk for one of the Hawaiian kings. There are many similar, smaller receptacles in the Museum in which the thongs are still in place.

Later an acquaintance of mine said that he had seen it and, without ever having read Admiral Rodman's article, described the "Sacred Calabash" in almost the same words as the Admiral so I think we must admit that the evidence is fairly good that this calabash actually existed and that the old Hawaiians might have been provided with a crude means of determining their latitude.

⚓

9

Off the Deep End

CHRISTOPHER MORLEY

SHE's what yachtsmen call one of the 12's; which means, I think, that she measures twelve metres on the water-line. But I won't be too sure about that, for the lingo of scientific yachting is full of conventional and arbitrary terms. As education for a philosopher I recommend a deep-water voyage in a racing craft on her maiden trip. For here is a beautiful plaything, a perfect theory, an algebraic equation of stresses (or guesses) and strains, existing previously in blue-prints only, suddenly put out to earn her first offing in the dirty weather of the Nova Scotia coast. I see her again, a white fancy in the opal shine of noon, as the tug *Togo* cast us off in the fog of Halifax Bay. New, untried, with stiff gear and 1952 square feet of canvas and all her brass winches still unverdigrised by salt.

So I won't be sure about her water-line, but her long beautiful over-hang, almost identical forward and aft, gives her 69½ feet over-all. If you lie on deck looking overstern (in gentle weather) and see how smoothly she slips through water, you'll perceive that she's more than mere theory. Afloat in a calm, under all her white canvas, she looks like a figure drifted from the pages of Euclid. Perhaps the idea is to make these racing craft as near an isosceles triangle as possible. Her tall mast (incredibly, terrifyingly tall to one accustomed only to knockabout craft: 80 feet above deck, 8 feet below) is stepped nearly amidship; and with marconi rig and a boom that does not project outboard you can imagine her an almost perfect segment of a huge circle. Her fore and back stays are the radii, her white hull the curve of the arc. To one all ignorant of racing boats everything about her in rig and gear was an astonishment. But certainly the internationalist finds her a good omen, for she was planned by a famous New York designer, built in Germany, her canvas is by Ratsey of Cowes and she was delivered in Halifax. *Iris* is her name.

It was in the *Nerissa,* during the two-day run to Halifax, that my spirit, always a lively foreboder, became aware of the fact that there is a great deal of water between Long Island Sound and Nova Scotia.

Iris, when we first saw her, together with *Tycoon* and several other German-built craft, was on the deck of the freighter *Lorain* which had arrived from Bremen only a few hours earlier. Securely frapped in cradles, they had made the voyage without mishap, but the hoisting them off by the big floating crane *Lord Kitchener* was an anxious business. That day it rained in a way that surprised even Halifax, a connoisseur of moisture. We stood about for hours in the downpour watching while the complicated job of unlashing and lifting the hulls was cleverly done. There was a curious eagerness in those two graceful shapes as the wire hawsers were gradually unbound. *Tycoon's* blue body, *Iris's* white, like pinioned gulls. They rose slowly, hung suspended from the crane, and were lowered overside. It was strange to see them come alive then. As *Tycoon,* unloaded first, was towed away, there was a sharp crack of thunder, almost like a salute.

By the time *Iris* was unloaded, after we had had a stout freighter's lunch of pea soup and corned beef and cabbage aboard the *Lorain,* the weather had cleared. *Iris* took the water without mishap. Riding a little high, without the weight of her big stick still to come, she dipped and swung gracefully. She knew her element. Now she was more than a blueprint.

So one loitered and watched our little tribe of argonauts make ready for sea. There were six in the flotilla: three 12's (*Iris, Tycoon, Isolde*), two 8's (*Whippet* and *Margaret F. IV*), and one very tiny cockleshell, the *Robin. Tycoon* had the outside berth, so we couldn't cast off until she did. *Iris* was ready; we had borrowed *Tycoon's* nail-clipper and all hands had trimmed their fingers, always the amateur's final gesture to civilization; not mere delicacy I assure you, but preparation for dealings with tough canvas. But still we must linger (to tell you the truth) because *Tycoon's* case of beer was late. So we lost those early airs from NW. It was towards noon before we got off. The weather was a warm hazy calm. We had to beg a tow from the tug *Togo,* to start us down the harbour. "Light Sly air" was the first entry in the log. "Sly" meant Southerly, but it might also have meant what it said. There was gentle insinuation in that weather and in the low barometer. Through milk-white banks of fog the *Togo* hauled us rapidly. She cast us off north of Neverfail Shoal. Our canvas was up. Now we were alone, we two, and could look at each other. Pearly haze thinned and thickened about us. We could see *Tycoon's* blue hull, with white waterline stripe and green underbody, leaping like a mackerel in the long swell. The high spires of canvas leaned amazingly upward; when the mist thickened we could not see the top. Running side by side we took stock of ourselves, tightened shrouds, compared chronometers. *Tycoon,* a tilting phantom of beauty, slid swiftly over the gray slopes. By her we could judge our own profile.

So with magical swiftness we were on our own. A tug, in a hurry to get back to another appointment, had rushed us down the harbour and cast us off—it seemed a little heartless—into a blanket of fog. Land was almost instantly out of sight, and our consort also. A long belly-wobbling sea came rolling under our bronze bottom. The chime of the Neverfail bellbuoy sounded like a summons to lunch, and from the cockpit one kept an eye on the swingtable in the main cabin. I had watched the stores going aboard. There, I said to myself, a large and frolic meal will be set out, such as yachtsmen enjoy. This was like old days in the ketch *Narcissus* where I myself had to do the cooking. There was a steward, seasoned by years at sea, to ration us. I thought (though a little dubiously) of the lobsters I had seen going aboard. But the corner of the table, visible from the cockpit, remained bare. No one said anything about food. I was much on my good behaviour. This was my first experience of real yachting. But, in the odd way one divines things, I felt that to say anything about food would (somehow) be amiss. I kept to looard of the Commodore, for I was taught young that one does not go to windward of the skipper. But his pipe (which, waking, he is never seen without) was very strong. Until about 2 o'clock I feared that perhaps there was not going to be any lunch. After that time my apprehension was different. I began to fear that perhaps there was. But about half past four (meanwhile nothing having been said) YG appeared with some slices of raisin bread. Then the truth came out. Our steward, the hardened seaman, was ill. We did not see him for four days.

Fog came down thick, and there was a steamer whistling not far away. She was inward bound round the lightship, we supposed; but the sound of her blast might have come from anywhere in an arc of nearly ninety degrees. A small fisherman's horn, pumped by hand, seems inadequate answer to that deep thuttering groan of a high-pressure steampipe. You get a very different sense of proportion when you hear a big ship's foghorn not from her own deck but from a small craft plunging from sea to sea. Suddenly the water seems very wrinkled and gray. Those waves are slate colour, even when broken they are not white but granite; they roll you in wet wastes of fog to teach you the blessings of being warm and steady.

The surprised faces of the lightship crew, as we passed close by them, might have suggested some surprise in our own minds. Our rig was evidently uncanny to them, and I was a little grim to remember how I had last seen that vessel, from the warm forward deck of the good old *Caronia* a level August morning. For now we were bundled up in all the half dozen layers of wool and oilskin, and chilly even so—always excepting the Commodore, exempt from all human weakness. And my testimony of the rest of that afternoon, as we zig-zagged (roughly speaking) SW and NW, must be, if honest, mostly of sleep. Such drowsiness as I have never known came down upon me. I fell loglike into a bunk and lay as one drugged and shanghaied. It was the miracle and quintessence of slumber, for one was dimly self-aware and knew how much one was enjoying it, yet too far gone for any shame or desire. One was as passive as a participle.

It was 7:30 on the morning of the first of June, so the notebook tells me, when we made Cape Sable abeam. The log said 212.5 miles and we reckoned the first leg of our cruise well accomplished. But only two hours later we found the mainsail parting from the brass slides that hold it to the track on the mast. So the mainsail was got down, and the trysail hoisted instead, while the skipper and Charley set about relashing the slides with wire. We then discovered that our patent log had somehow chafed through and gone adrift. Thereafter the Commodore reckoned our speed by throwing an empty matchbox overboard at the bow and timing it to the stern with a stop-watch.

That night there were mares' tails in the sky, long skeins and streamers of cloud brightened by the moon. By 6:30 a.m. sea and wind were rising merrily. There was no talk now of putting on the mainsail. Even on the trysail the lashings of the slides were beginning to go, she was taken down and reefed. There came pouring rain and a strong SW gale. The jib also we took down. Now, unless we ran with a bare pole, this was all that could be done. The glass between 29.50 and 29.55.

When you speak to me of the Bay of Fundy, that is the day I shall remember. When one was below, the morale was not too good. This was now the fourth day, and what with one thing and another the cabin had not had a cleaning since we sailed. The sea had been rough and those not on actual duty had no ambition for anything but sleep. The patent German ash-tray, come from Bremen, had capsized first of all and spilled matches and tobacco everywhere. Water coming liberally through the skylights had moistened everything to a paste. *Iris,* leaping merrily among hills and valleys, was easing herself to the strain, but her chorus of creaks and groans was anxious to those below. Large consignments of ocean came upon her with the heavy solidity of an automobile smash. How wet were those brown blankets! I admit that the chronicler and YG, brooding below and watching cracks widen in the bulkheads and panellings, had a vague notion that she might dissolve about them. The Commodore, coming down to examine the chart, was entertained to find his underlings suggesting it would be a good time to seek shelter somewhere. He was quite right, of course; we were best where we were.

But above, when one's eye grew accustomed to the size of that sea and the way she handled herself, there was real thrill. How big were the waves, people always ask? It cannot be answered because in a heavy sea the hills are too broad to allow the eye any fair scale of measurement. But you see them with a different eye from that of the passenger of a big liner. On a big ship you look down on the water and its colour seems darker. From *Iris* we looked closely into those long ridges that loomed above; we could see how coldly green and translucent they were. Every once in so often there was some particularly big comber one could mark from far away: it came striding, breaking in a crest a hundred yards long, with a definite menace written all over it. There was something unpleasantly personal about those waves. "I'll get you if I can," seemed to be their autograph. They would rise, perhaps thirty feet above us, leaving us momently in a dull green twilight, far down the hollow. Then with the soar of a rising gull, she would ride up as the great shoulder lifted her. A swirl of cream about her nose as the comber spilled a few buckets along her deck, and we gazed triumphant from the summit along leagues of water laced and wrinkled with foam. For nine or ten hours we were practically hove to, riding switchback on these big ones. Wind sang in our rigging, rafts of fog swathed us in. It was a specially big sea coming through the skylight late in the afternoon that really brought us round the corner. Several gallons of cold water soused on the Commodore's head as he lay asleep. He sat up promptly, looked about at the foul mess in his pretty cabin, and remarked only, "Well, boys, let's clean up." Somewhat gingerly, creeping about in that frolicking hull, we did so.

Somewhere in those waters, perhaps still faintly perfumed by the Commodore's pipe, there is an invisible longitude, a Shadow Line, where the Bay of Fundy becomes the Gulf of Maine. For when the Commodore roused his starboard watch at 4:14—having given them an hour and a quarter as lagniappe—there was that good feeling of having turned a corner unawares, some unseen facet of space and time. Now, with gales and chilblains left astern, was time to resume the famous mainsail. At 6:15 we took down the trysail and were ready to raise our full-page spread. I particularize the episode because it is a parable of the uses of indolence. The chronicler, always evasive of toil, was wont to give an apparent demonstration of zeal, to justify himself in the Commodore's eye. When great weights of canvas were to be handled or hoisted he could cry *hoick* (or however you prefer to spell that rhythmical groan) with the loudest, but it was mostly subterfuge. And now, while Captain Barr and Charley and the PR were lustily tailing onto halliards and winch, the chronicler was standing by (keeping that big mast between him and the Commodore) and sojering. He was pretending to be doing something, I don't know just what, but in reality he was surveying all that intricate gear with his usual questioning amazement. So it was he who observed that the bronze gooseneck, which holds the boom to the mast, was cracked almost through. The metal had gone a sort of roquefort cheese colour and was radiated with fissures. Obviously the thing was unsafe. In that pleasant breeze and with so gross a canvas the thing would most likely snap, there would be a big boom thrashing loose and all sorts of devilment.

With reluctant hearts the company abandoned the proud mainsail for good and all. The boom was unshipped and lashed on deck. Up again went old standby, the topsail. And the mainsail was stowed in the Commodore's cabin where it filled all the space and where that uncomplaining commander crept in and out of his bunk like a chipmunk in a tangle of underbrush. This was a two-hour job. Now, in the first clear sunshine and fair breeze of the voyage, we must go soberly along under storm gear. *C. G.* 24, a smart destroyerish lady in naval gray, passed near us and evidently took note of our cautious demeanour. We had, to sea eyes, much the look of a man who attended a smart wedding in cutaway garb and a golf cap. But there is one great etiquette among ships: you know there is a valid reason for everything, and don't ask rude questions.

That sunny forenoon was notable first for a series of sextant observations. At high noon, after a final flurry of sun-shooting, the two navigators gave out statements. Our position was authoritatively stated as 42° 54' N., 69° 47' W. You can look it up on the chart and see how near it tallied with naval prognostics. This reckoning was entered in the log as our position at Wedding Time. For it so happened that this date was the marriage anniversary of one of the company. *C. G.* 24 was no longer in sight, and anyhow we were not yet in territorial waters. The sound of splitting wood was heard from the main cabin, where we had been barking our shins on those cases for four days. The other three, hearing the Commodore hatcheting down below, looked at each other with a sweet surmise. An empty box came flying up through the companionway and fell with an agreeable thump onto the calm ripple that slid softly by. Honourable men know how to solemnize a date of sentiment. Need I insist what was the stencil on those floating jetsams? Mumm's the word.

We drifted W by N into a warm sunset. Still perhaps unduly doubtful of our reckonings, we drew lots as to what part of the coast we would pick up. At 6:56 p.m. we sighted Cape Ann on the port bow. But we were quite helpless in that delicious calm. We swam idly with the tide. In the dusk we sighted a small power schooner lying enigmatically offshore. With foghorn and flashlight we hallooed her, thinking perhaps she would give us a tow to Rockport; but she paid no heed to our signals. When we drifted alongside she replied to our hail with inhospitable monosyllables.

Marblehead, the font and chrism of New England yachtsmen, was an old story to those others; but all new to the simple chronicler. He rose through the hatch, about six o'clock of that blithe airy morning, and found *Iris* rippling through a strait of bouldered coves. One with several good hours of oblivion behind him looks dispassionately on the vigils of others, so he forbore to chaff the Commodore on the amount of to-bacco ashes sprinkled along the cockpit coaming, spoor of the com-mander's all-night watch. "You can take the jib in," were the Com-modore's exact words at 6:18, in a tone worthy of Cabot. There was more than just the due severity of great commanders in this long vigil of his. I think he had wished to spare his new ship the embarrassment of publicly arriving in so tony a harbour under jury rig. As Emily Post, if her stocking should choose to run, so (I divine) would *Iris* have felt to appear under trysail opposite the verandah of the Eastern Yacht Club. So, in the morning hush, while even the lobsters destined for Marble-head's luncheon were hardly alert, we stole in among many handsome craft and let go our hook. Four days 19 hours 35 minutes we reckoned our passage.

Our consort the *Duenna,* a stout and sea-kindly power boat, met us at Marblehead. The wind was light and contrary, so she took us in tow. There were still three days, but they were of a quite different psychology and require no special exegesis. Except when the *Duenna* (admirably named, for she chaperoned us as though *Iris's* virtue were of the frailest) left us to dart into various ports for fuel, we made the rest of the voyage at the end of a line; and a ship in tow is a mere somnambulist. This is not to say there were not many pleasures to ponder. I shall not be ex-cessive on this topic, but the last two days were very largely concerned with grub. We had four or five empty days to catch up with. Now that there was little navigating to consider, almost at any moment some member of the company could be found eating. The PR discovered some marmalade among the stores, and remained at table long after all others. "I haven't been eating enough marmalade in the last three or four years," was his excuse.

So with the agile *Duenna* running ahead of us like a cottontail rabbit, kicking up a plume of spray and exhaust, we came swift along. We picked up the familiar landmarks of the Sound one by one, including the two desolate masts in Fisher's Island Sound. They testify tragedy where some shipmaster missed the channel by only some 50 yards. In-land waters give one plenteous parable. The habituated landsman thinks of large bodies of sea water as a liquid subject to embarrassing up-and-down movement, but fairly stable on its base. But—as the Bay of Fundy or Long Island itself will promptly tutor you—these vast masses are excessively fluid and move to and fro in the most surprising fashion.

A little study of Current Tables and Tide Diagrams is highly illuminating; or a glimpse of the tide boiling through the Race, between New London and New Haven. And why, I've often wondered, does the Coast Guard get so little acclaim for its quiet, faithful and endless work in keeping buoys, beacons and bells in constant A1 service? Consider the lives and property daily and nightly confided unquestioning to these safeguards we all take for granted as we do the phenomena of stars and weather. An occasional halloo of gratitude would not be amiss.

⚓

10

The Story of Sailing

JAMES THURBER

PEOPLE who visit you in Bermuda are likely to notice, even before they notice the flowers of the island, the scores of sailing craft which fleck the harbours and the ocean round about. Furthermore, they are likely to ask you about the ships before they ask you about the flowers and this, at least in my own case, is unfortunate, because although I know practically nothing about flowers I know ten times as much about flowers as I know about ships. Or at any rate I did before I began to study up on the subject. Now I feel that I am pretty well qualified to hold my own in any average discussion of rigging.

I began to brush up on the mysteries of sailing a boat after an unfortunate evening when a lady who sat next to me at dinner turned to me and said, "Do you reef in your gaff-topsails when you are close-hauled or do you let go the mizzen-top-bowlines and cross-jack-braces?" She took me for a sailor and not a landlubber and of course I hadn't the slightest idea what she was talking about.

One reason for this was that none of the principal words (except "reef") used in the sentence I have quoted is pronounced the way it is spelled: "gaff-topsails" is pronounced "gassles," "close-hauled" is pronounced "cold," "mizzen-top-bowlines" is pronounced "mittens," and "cross-jack-braces" is pronounced "crabapples" or something that sounds a whole lot like that. Thus what the lady really said to me was, "Do you reef in your gassles when you are cold or do you let go the mittens and crabapples?" Many a visitor who is asked such a question takes the first ship back home, and it is for these embarrassed gentlemen that I am going to explain briefly the history and terminology of sailing.

In the first place, there is no doubt but that the rigging of the modern

Reprinted by permission of *The Bermudian*.

sailing ship has become complicated beyond all necessity. If you want proof of this you have only to look up the word "rigging" in the Encyclopedia Britannica. You will find a drawing of a full-rigged modern ship and under it an explanation of its various spars, masts, sails, etc. There are forty-five different major parts, beginning with "bowsprit" and going on up to "davit topping-lifts." Included in between are, among others, these items: the fore-top-mast staysail halliards (pron. "fazzles"), the topgallant mast-yard-and-lift (pron. "toft"), the mizzen-topgallant-braces (pron. "mazes"), and the fore-topmast backstays and topsail tye (pron. "frassantossle"). The tendency of the average landlubber who studies this diagram for five minutes is to turn to "Sanskrit" in the encyclopedia and study up on that instead, but only a coward would do that. It is possible to get something out of the article on rigging if you keep at it long enough.

Let us creep up on the formidable modern sailing ship in our stocking feet, beginning with one of the simplest of all known sailing craft, the Norse Herring Boat. Now when the Norse built their sailing boats they had only one idea in mind: to catch herring. They were pretty busy men, always a trifle chilly, and they had neither the time nor the inclination to sit around on the cold decks of their ships trying to figure out all the different kinds of ropes, spars, and sails that might be hung on their masts. Each ship had, as a matter of fact, only one mast. Near the top of it was a crosspiece of wood and on that was hung one simple square sail, no more complicated than the awning of a cigar store. A rope was attached to each end of the cross-piece and the other ends of these ropes were held by the helmsman. By manipulating the ropes he could make the ship go ahead, turn right, or turn left. It was practically impossible to make it turn around, to be sure, and that is the reason the Norsemen went straight on and discovered America, thus proving that it isn't really necessary to turn around.

As the years went on and the younger generations of Norsemen became, like all younger generations, less hardworking and more restless than their forebears, they began to think less about catching herring and more about monkeying with the sails of their ships. One of these restless young Norsemen one day lengthened the mast of his ship, put up another crosspiece about six feet above the first one, and hung another but smaller sail on this new crosspiece, or spar (pronounced, strange as it may seem, "spar"). Thus was the main topsail born.

After that, innovations in sails followed so fast that the herring boat became a veritable shambles of canvas. A Norseman named Leif the Sailmaker added a second mast to his ship, just in front of the first one, and thus the foremast came into being and with it the fore mainsail and the fore topsail. A Turk named Skvar added a third mast and called it the mizzen. Not to be outdone, a Muscovite named Amir put up a third spar on each of his masts; Skvar put up a fourth; Amir replied with a fifth; Skvar came back with a sixth, and so it went, resulting in the topgallant foresail, the top-topgallant mizzen sail, the top-top-topgallant main topsail, and the tip-top-topgallant-gallant mainsail (pron. "twee twee twee twa twa").

Practically nobody today sails a full-rigged seven-masted ship so that it would not be especially helpful to describe in detail all the thousands of different gaffs, sprits, queeps, weems, lugs, miggets, loords (spelled "leewards"), gessels, grommets, etc., on such a ship. I shall therefore devote what space I have left to a discussion of how to come back alive from a pleasant sail in the ordinary 20- or 30-foot sailing craft such as you are likely to be "taken for a ride" in down here in Bermuda. This type of so-called pleasure ship is not only given to riding on its side, due to coming about without the helmsman's volition (spelled "jibe" and pronounced "look out, here we go again!"), but it is made extremely perilous by what is known as the flying jib, or boom.

The boom is worse than the gaff for some people can stand the gaff (hence the common expression "he can stand the gaff") but nobody can stand the boom when it aims one at him from the floor. With the disappearance of the Norse herring fisherman and the advent of the modern pleasure craft sailor, the boom became longer and heavier and faster. Helmsmen will tell you that they keep swinging the boom across the deck of the ship in order to take advantage of the wind but after weeks of observation it is my opinion that they do it to take advantage of the passengers. The only way to avoid the boom and have any safety at all while sailing is to lie flat on your stomach in the bottom of the ship. This is very uncomfortable on account of the hard boards and because you can't see a thing, but it is the one sure way I know of to go sailing and come back on the boat and not be washed up in the surf. I recommend the posture highly, but not as highly as I recommend the bicycle. My sailing adventures in Bermuda have made me appreciate for the first time the essential wonder of the simple, boomless bicycle.

11

A Narrow Escape in the Fastnet Race

IRVING JOHNSON

THE Fastnet Ocean Race is sailed over what is known as the toughest ocean race course in the world. The contestants have to sail down the English Channel and past Lands End, and then across the Irish Sea to Fastnet Rock, about five miles off the southern coast of Ireland. There is a lighthouse on the rock, and that enables the racing yachts to round the rock at night, a very important advantage. The lighthouse and the keeper's house together cover almost the entire top of the rock, which rises about thirty feet above sea level. In bad storms the waves go right over the keeper's house and half way up the lighthouse. The race continues back around Lands End, past the Lizard, and up the English Channel to the finish line at Plymouth, making a total distance of six hundred miles.

One of the ambitions of every deep sea yachtsman is to participate in the Fastnet Race. It gives him prestige so that to some degree he can look down with a feeling of superiority on yachtsmen who haven't had the opportunity. I was no exception in my sentiment about the Fastnet Race, and I went aboard one of the trans-Atlantic racers owned by an American, and talked with him about the possibility of sailing on his boat. He said he would like to have me go along as a member of the crew, but I didn't like the looks of him or his rigging, and decided to try elsewhere.

Back I went to the *Wander Bird,* and the skipper appealed successfully in my behalf to another American, George Roosevelt, whose yacht, the *Mistress,* was one of the swiftest in the fleet. He himself has taken a part in the computing of some complicated navigation tables, and is considered one of the ablest amateur navigators.

The race was to begin at noon. As the time approached, the wind went down, and when the starting gun boomed at exactly twelve o'clock, there was a loud splashing of anchors being thrown over the side amongst the racing fleet. All the skippers had let their boats drift down with the swift current, and many of the craft had crossed the starting line going backward just about the time the gun was fired. The *Mistress* was going backward so fast that one anchor wouldn't stop us. So after dragging down through the fleet we had to throw over our largest anchor in order to bring the yacht to a standstill.

We noticed that two English boats, which we knew had radios aboard, were heading more to the southward toward France than the rest of us. It seemed to the skipper and mate of the *Mistress* that probably those two boats received weather reports, and that those sailing them knew more about the weather to be expected in the English Channel than any of their rivals. So we also headed southward, figuring that they expected a southerly wind the next day. But about the time we sighted the coast of France the wind gradually came around to the northward, which was just the opposite from what we looked for. This put us in a bad position right at the start of the race. When we finally got out to Lands End the wind shifted around behind us and blew harder and harder as night came on. We sailed into the Irish Sea at a fine clip, but about eight hours behind where we would have been had we kept along with the other boats.

The spinnaker was set on one side, while the mains'l was slacked far out over on the other side, and we ran dead before the wind faster and faster as the night wore on. Presently Mr. Roosevelt, who was both owner and skipper, and responsible for the whole ship, asked Mr. Hoyt if he didn't think the spinnaker ought to be taken in, the wind was blowing so hard. After a moment Hoyt replied: "Well, no. Let's drag it a little longer and see what happens."

The night got blacker, and the wind still increased. A half hour passed. Then Hoyt moved over nearer to Roosevelt so he could make him hear, and called out, "Say, George, if we're going to save that spinnaker in one piece, we'd better take it in!" But by that time the skipper himself had decided to let it stay for a while longer.

A half hour of this passed, and the skipper said, "Hoyt, we ought to take that spinnaker in now." But Hoyt's response was, "I think that considering it has stayed up there so long it might remain a little longer."

So stay it did for another half hour period, when essentially the same conversation was repeated, and the spinnaker stayed aloft. By this time it was apparent to both men that the wind was so strong the straining sail couldn't be taken in without carrying away something, and they decided that the wind might as well take care of it altogether. A little after midnight we saw numbers of lights ahead on both sides, and we knew that these must be on some large steam trawlers from England that were out there dragging their nets across the shallower parts of the Irish Sea. As we rapidly got closer we saw that one of them would just about hit us if we both kept on at the same speed. But there was nothing we could do about it, because we were running dead before the wind.

Apparently the trawler didn't see our side lights. On getting nearer we directed our flashlights on the trawler, but still there were no signs of life. I was busy with one of the flashlights, and things looked to me as if we certainly would be run down. Then suddenly just above my head came an explosion, and a great ball of fire blew out ahead of us right over the bridge of the trawler. From our boat the skipper, standing just back of me, had fired a Very pistol. The fearsome ball of fire which came banging out of the black night surely must have awakened the fisherman's crew, but we were too close and going too fast for them to help us now.

The massive steel sides of the trawler loomed up in the light of the distress signal, and we saw that she must have a tonnage ten times that of ours. In our minds we could visualize just how the *Mistress* would splinter and crumple up against that dripping black hull in the next few seconds. Then an extra strong puff of wind hit us, and a wave lifted our stern and pushed us ahead faster than ever, so that with one grand sweep down the side of that wave we just skimmed past the bow of the trawler. The end of our spinnaker pole missed her by only a few feet, and as her bow reared up close to our stern the trawler's lookout stuck his head over the bow and yelled out, "Hey, your lights ain't very bright, Sar!"

The next morning the wind eased off some and hauled more ahead. So we took in the spinnaker, and about midday rounded Fastnet Rock.

Before this we had noticed several racers that had outsailed us but now their lead was cut to about three hours. This meant that by carrying on all night we had gained valuable time, and we set out after them back across the Irish Sea. When we neared Lizard Point on the way home the waves evidently were converging to their utmost, while a strong breeze on the quarter was pushing us through the sharp seas at our maximum speed of about eleven knots. At that time we were carrying a mains'l, ballooner, and a balloon maintopmast stays'l, which Sherman Hoyt had nicknamed the "golwobbler."

As we rushed into that rough, choppy sea off the Lizard, the boat jerked and twisted and half buried herself time after time. This was such a fine sight that I decided to make pictures from the mast. Just as I swung my foot over the spreader, the large iron hook of the back-stay tackle straightened out under the strain of the golwobbler, and with a twang the backstay flew up in the air and around to the back side of the mains'l. I slid down the mast so fast I must have broken the sliding record for I reached the deck before the men in the cockpit aft had run forward to lower the golwobbler.

It came down in a hurry, but this cut our speed down to two or three knots, which was too tantalizing to be endured, especially as our competitors whom we had been gradually gaining on, were in sight up ahead and now were walking right away from us. If we could only get hold of that loose backstay, dangling out there on the far side of the mains'l, we could rig up another tackle and set full sail again.

From my experience in a square-rigger, I felt confident that I could get hold of the stay by going out under the main boom. The wind was blowing too hard to luff with such an amount of sail on, and that was our only chance. So by hanging on to the main boom while my body dangled underneath, I gradually worked my way to the end which was far out over the water. As the wire stay came swinging around, I grabbed it in my teeth and worked back along the boom the way I had come, but on the windward side where the stay was to be fastened again. Soon we had full sail set and were roaring on. This little stunt of handling the stay under such conditions was written up in practically all the yachting magazines in England and America. But really it wasn't very much for a fellow used to the yards of a square-rigger.

We bowled along into Plymouth Harbor and crossed the finish line before nightfall. This was on the fifth day of the race, and we learned that we had taken third place among the seventeen yachts competing. If we had not made that side run off to France we would have been the first boat back, and probably would have had second place on time allowance.

⚓

Fifty South to Fifty South

WARWICK M. TOMPKINS

A man who loves the sea and ships can aspire to no more searching test than a Horn passage. It is the last word in the lexicon of sailormen. There nature has arranged trials and tribulations so ingeniously that in the van of all synonyms for sea cruelty and hardship is the ironbound name of Cape Horn. Winds blow elsewhere at times as strongly as they do south of fifty. Seas elsewhere may pyramid as high, break as heavily. There may be places equally remote and as bleakly lonely. Currents in other regions may be as adverse. These foes the sailorman may encounter separately or in pairs here and there, aye, encounter and best, but always in his heart he will wonder if he could face all combined. If he glories in the unequal contest of human muscles and artifices with the ocean, if the sea shouts an insistent challenge, he can never be truly content until he has voyaged from fifty south in the Atlantic to fifty south in the Pacific in his own command. This is the ultimate test, given to very few to know.

YET again wind and sea snatch brief midnight rest. Just so seven of the last nine days have been ominously stillborn. The early dawn of the southern spring finds the ship still north and miles east of the Horn, victim of eddying current, leeway and this new calm.

Refreshed and gleeful at new-planned deviltries the wind comes whistling from the northwest at three. It feints to the northward two full points and we are lured into setting the full mainsail. With eased sheets the *'Bird* romps for the Horn.

"Breakfast in the Pacific!" forecasts John. "We knocked off eight knots this last hour."

Close aboard the terminal ranges plunge sharply into the ocean. Anson named them Tierra del Fuego, Land of Fire. They have altered not at all since he, with loathing and dread, thus christened them. Blackly-blue, rent by deep, mysterious canyons, crowned with snow these mountains (contrary to popular opinion) know nothing of volcanic pyrotechnics or vaporing geysers. Even as it is today it has always been; not a single wavering smoky spiral climbs into the solid gray clouds.

And so long as Anson held his luck he saw no fires, either. They sprang up, leaping sparks blown from peak to peak, only when he started losing his ships and men on the plentiful dangers hereabout. Then each brief, bright signal rallied to slaughter and looting the hardy primitives who, for reasons beyond comprehension, are native to this region of eternal unkindness.

Such quick, brief flames may well leap again, and for us, if we but touch the coast where sharp, peering eyes can appreciate our helplessness.

Missionary efforts have modified but slightly the barbarically simple manners of the Tierra del Fuegians. Our new charts print warnings to those trafficking with them; only twenty years ago they baldly stated these people were cannibals. Captain Joshua Slocum's most enduring bid to fame was made when he shrewdly planted his decks with big carpet tacks, points up, and effectively kept *Spray* clear of bare-footed, murderous visitors in the Straits of Magellan.

Barr, our anthropologist, tells of these people whose habitations are mere lean-tos fronting their campfires, who wear only the most rudimentary fur garments, and those with the fur outside. Their food is the meat of seals and the flesh of fish, stewed with edible sea plants and the local wild celery which has in times past saved so many mariners from scurvy.

Luffing sails abruptly terminate Fuegian speculations. The north wind shifts eight points. It puffs just once from the west and then drops to nothing. With her sails all emptied the schooner glides for a space, borne by her momentum. Before she can come to rest the unnaturally servile sea, aping the masterful wind, has itself settled into treacherous repose.

Entr'acte!

Our very cigarette smoke hangs heavily inert. The wind that is gone—

or is it the wind which is to come, or perhaps both in duet?—whines faintly. From the swiftly altering gigantic stage we get our imperative cue and are sprung to sudden labor.

"Double reef that mainsail, and quickly!"

The spectacle is consummate theater, colossal art enlisting a sky, a sea from horizon to horizon, forces incalculable, properties and effects monstrous, awful and dreadfully real.

Two-thirds of our universe are yet unchanged, all monotonously solid gray. The shores we have skirted and left behind remain sharp, blue cameos.

But see! The West!

From north to south, arched, boiling, oily and solid, is a line squall momentarily higher and closer. Under it sways an impenetrable leaden pall whose fringe tickles and darkens the sea. Its folds sweep over the Horn, hanging it with gauze through which we can see the mountain-cape but vaguely. It is, therefore, less than seven miles away, and coming with hurricane speed. Already catpaws stroke the water and fan coolly on our cheeks.

"Tack lashing all fast!" William yells. "Stretch your sail!" The precision with which everyone works is gratifying.

"Vast heaving! Tie in your reef!" I pass the clew lashing, heaving each turn taut.

A glance shows the storm now very near; the keystone of the arch is in our zenith. The swift interplay and convolutions of its clouds surpass anything I have ever seen. The windward sea shows white teeth. About us fleet ripples scurry away into the east. The air is atremble with the rising, whistling wind.

At this instant I discover the minute tear in the tablings of the clew. It is an insignificant little rip, only an inch in length. Five minutes with needle and thread will mend it.

"Down mainsail! Get her in!" A stitch in time is worth far more here than in pleasanter climes.

We jump fast, hoping to get the loose sail captured before the wind arrives, but the squall strikes first. This wind is more than just wind; it is a solid, palpable bludgeon. Its blow staggers the ship.

The mainsail is all but in when there is a blasting rasp of tearing cloth. This all-pervasive violence has unerringly found the sail's insignificant weakness. Cotton is shredded down-wind and the quick scream of split canvas is in our ears. We cannot even glance at this wound. Other sails must be got in at once, and oil out before we are stricken more vitally.

This is our most furious blow thus far. There is no estimating accurately the wind's force. It is gale piled upon gale, and no matter of miles per hour can convey a sense of turmoil. We log it at force 10, but we are growing hardened to severe weather and it is quite probable familiarity is breeding not contempt but an objective power of understatement and depreciation which should properly go with wide experience. The wildest sea stories are always told by those who know least of the sea.

Is this gale a "heavy gale"—to use Admiral Beaufort's notation—or a "very great gale, a tempest"? He assigns to them a difference of only fourteen miles an hour. It is quite possibly a hurricane. Between a tempest and a hurricane Beaufort put a mere seven miles an hour. I'd like to know more about his arbitrary limitations, the way he measured his velocities, and if he allowed for the anaesthetizing effect of experience.

To bring exactly what I mean closer to the landsman let him consider how fast fifty miles an hour seems on the open road after passing through a town at thirty. How impatiently he trails a truck at the same fifty after an hour spent at sixty or seventy!

It is just so with gales. I have often noted before, as my crew is discovering now, that when one lives incessantly with heavy weather a forty-mile blow can seem relatively mild and pleasant. After a ship has gone far to leeward, hove-to and beset by foul current, a two-knot crawl under the closest storm sails seems by comparison very fine sailing.

In this kind of weather the mainsail cannot be carried even if it were not damaged, so we are little concerned. The storm trysail is more easily handled and more comfortable. It is no weather to be sewing on deck, and we complacently count on postponing repairs until we are north of Fifty.

Meanwhile it is made very apparent that we are going to have to sail, sail, sail if we want to get to the northwest. Consequently, the first fury of the gale having abated somewhat, we set the jumbo again at 9:30 this morning. Our rig is fantastically short—deeply reefed foresail and the jumbo—but hereafter we shall often count ourselves lucky when we can carry this much canvas. We shall come to congratulate ourselves when we can also set the inner jib, and it will be a joyous occasion when the outer jib is hoisted. Hallelujahs will hail the tiny storm trysail on the main mast.

Noon Position: 56°—19′ S., 67°—10′ W. (D.R.)
To 50° South in the Pacific: 603 Miles.
Made Good: 8 Miles. Sailed by Log: 68¾ Miles.

The jumbo is set now for less than an hour. The gale rages with new strength again, but the seas are lengthening and treating us less severely. At noon we are almost into the Pacific. Although thirty-three miles from yesterday's position, having sailed nearly seventy, the ship is only eight miles nearer the fiftieth parallel since most of our progress has been to the southward.

It blows all afternoon, but at suppertime we again get moving with the help of the jumbo. At eight o'clock we set the storm trysail.

By midnight it is again calm. What a place!

⚓

13

The Classification of Racing Yachts

WILLIAM H. TAYLOR

To THE yachtsman who is accustomed to speaking of and dealing with boats in the various racing classes, it comes as something of a shock to hear a landsman ask earnestly, "What's the difference between a Six-Metre and a Class J boat, and why isn't a K boat bigger than a J?" But a brief academic consideration of the subject, with its various ramifications and contradictions, will put the yachtsman in sympathy with the landsman's confusion.

In fact, he realizes that the confusion is no more unnatural than his own when he tries to define the exact nature of a claiming race, a futurity, a furlong, an offside, a dedans, an error, a niblick, casual water or a balk-line. Nobody can be expected to understand the technical meaning of every term in sports he doesn't take part in, but frequently one likes to have a general idea what his friends and the papers are talking about, hence explanations may be in order.

The classification of racing yachts is complicated by the fact that they are arranged in four parallel sets of classes, which leads to the peculiar situation that a yacht may have from one to three or four different ratings, depending on what sort of a race she's in. The four general types of classification in common use are the Universal Rule, the International Rule, the Cruising Club (or ocean racing) rule and the one-design classes, and it is quite possible for a New York forty-footer, which started life as one of a one-design class, to race at one time or another under each of the three rating rules, with a different rating and correspondingly different time allowance. (This point is brought out not so much to confuse the reader as to ease him gently into the spirit of further complications which follow.)

Reprinted by permission of the *New York Herald Tribune*.

TABLE OF PRINCIPAL DIMENSIONS FOUND IN YACHTS
OF FOUR WELL-KNOWN CLASSES

UNIVERSAL RULE CLASSES

Class (Boat)	L.O.A. Ft.	In.	L.W.L. Ft.	In.	Beam Ft.	In.	Draft Ft.	In.	Sail Area	Crew
76-rating (J) (*Rainbow*) ..126		7	82	0	20	11	14	11	7,572	26
46-rating (M) (*Valiant*) .. 80		9	54	0	14	0	10	4	2,739	10
25-rating (Q) (*Robin*) ... 49		0	32	0	9	0	6	10	886	5–6
20-rating (R) (*Robin*) ... 40		2	26	9	7	0	6	0	588	5

INTERNATIONAL RULE CLASSES

Class (Boat)	L.O.A. Ft.	In.	L.W.L. Ft.	In.	Beam Ft.	In.	Draft Ft.	In.	Sail Area	Crew
12-Metre (*Cantitoe*) 68		9	43	11	12	4	8	8	2,050	8
8-Metre (*Priscilla*) 48		0	30	0	8	3	6	3	854	5–6
6-Metre (*Challenge*) 37		0	23	5	6	4	5	4	451	4–5

CRUISING CLUB RULE (INDIVIDUAL BOATS)

Yacht (Rig)	L.O.A. Ft.	In.	L.W.L. Ft.	In.	Beam Ft.	In.	Draft Ft.	In.	Sail Area	Crew
Duckling (Sloop) 37		7	29	6	10	6	6	0	630	5
Dorade (Yawl) 52		2	37	0	10	3	7	8	1,150	7
Teragram (Schooner) 58		5	43	2	14	3	7	10	1,843	10
Mistress (Schooner) 60		6	50	0	15	7	9	8	2,300	11
Vamarie (Ketch) 70		2	54	0	15	3	10	4	2,200	12

ONE-DESIGN CLASSES

Class (Boat)	L.O.A. Ft.	In.	L.W.L. Ft.	In.	Beam Ft.	In.	Draft Ft.	In.	Sail Area	Crew
12-Metre O.D. 69		2	43	11	12	8	8	5	1,976	8–9
10-Metre O.D. 58		10	35	0	10	6	7	6	1,432	6–7
32-ft. 45		4	32	0	10	7	6	6	950	5–6
30-ft. 43		6	30	0	8	10	6	3	1,103	5–6
Victory 31		6	20	8	7	0	4	10	450	4–5
Interclub 29		6	19	0	7	10	4	6	415	4–5
Atlantic 30		6	21	6	6	6	4	9	383	3–4
Star 22		7	15	6	5	8	3	4	280	2
Snipe 15		6	—	—	5	0	—	—	117	1–2
Dinghy (Class D) 10		0	—	—	4	0	—	—	67	1–2

Explanation of table—L.O.A. is over all length. L.W.L. is length on waterline. Sail area, in square feet, includes working sails only, spinnakers and balloon jibs not measured. Crew here is approximate number of men ordinarily carried in racing, including both professionals and amateurs, and in case of Cruising Club Rule boats provides for two watches as is customary at sea. All yachts except 20-raters, 6-Metres, and one-design classes from Victories down, have living quarters aboard for both professionals and afterguard. Most of the Cruising Club Rule boats, and the one-design, 12-Metres, 32-footers and some of the 10-Metres, have auxiliary motors. Unless otherwise specified, rigs are sloop.

The Universal, International and Cruising Club Rules are all mathematical means of obtaining ratings. The ratings, expressed in feet or metres, bear no fixed relation to any one dimension of the boat, although in a general way it is usually somewhat less than the waterline length.

From this rating the yacht's handicap (or time allowance, to be exact) is determined—the number of seconds a mile she gets from a real or imaginary scratch boat. This is used to translate the boat's elapsed time in any race into her corrected time, on the basis of which the prizes are awarded in handicap racing.

In non-handicap, or class, racing each yacht within a given range of rating length is assumed to be at the top rating of that range, and neither gets nor gives time, hence racing yachts generally are built to the top rating of some one class under either the Universal or International Rule. (This doesn't apply to ocean racing boats rated under the Cruising Club Rule.)

The rating in any of these three rules does not correspond to any given dimension of the yacht, but is a figure obtained by an equation which takes into consideration several different measurements. For instance, the rating of a Universal Rule yacht is "18 per cent of the product of the length multiplied by the square root of the sail area divided by the cube root of the displacement (weight to you) of the yacht." Length, incidentally, is no one simple dimension but is obtained by a formula in which the waterline length and the quarter-beam length (length on a certain plane in the hull to one side of and above the load waterline at the keel) are both considered. There are other limiting and contributory factors.

This rule is quoted because it is the simplest of the three. Some of them are really complicated, but all, eventually, produce a rating. Any one desiring to pursue the subject further is referred to any one of the excellent technical schools of naval architecture. Even some professional designers occasionally seem to find themselves baffled by these rules, and the beginner is strongly advised to take our word for it that if he owns a boat, and has her measured by an accredited measurer, and turns his measurement certificate over to his club officials, he will eventually get a rating and a handicap which will probably be the right ones, unless the club secretary gets the mail in the wrong envelopes.

Having thus elucidated the whole problem of rating rules, we turn to the practical application thereof. The Universal Rule, developed many years ago in this country by N. G. Herreshoff and the New York Yacht Club, and commonly used by that and other leading clubs in their regattas and port-to-port races, contains time allowance tables for yachts of from 15 to 150 feet rating, but actually the important classes racing under it in recent years have been the twenty-raters (Class R), twenty-five raters (Class Q), forty-six raters (Class M) and seventy-six raters (Class J).

In mixed events, such as the New York and Eastern Yacht Club cruises, where all sizes of yachts race together for handicap prizes, other individual yachts, and yachts built to other classes, are measured and handicapped in whatever class they happen to fall in. (Each class is designated by a letter merely for convenience, starting the alphabet with the largest boats.)

The International Rule, which is used throughout Europe and the British Isles and which is also used here, owing to our participation in international racing, works similarly, though somewhat different factors are considered, such as sail area and freeboard. The important inter-national rule classes here are the Six-Metre, Eight-Metre and Twelve-Metre classes, whose respective ratings are just what it says, if you can figure in metres. On events like the club cruises the "metre-boats" generally race under a Universal Rule rating. For instance, the Twelve-Metres rate down at the bottom of Class M, being somewhat smaller than the M boats, and the Eight-Metres are pretty close to the Q's in size. But whereas all the Twelve-Metres, for instance, rate the same under the International Rule, their ratings and allowances vary under the Universal Rule and in general they fare somewhat worse than the boats built specifically to that rule.

Except in the cruises mentioned above, handicap racing has waned in popularity and a boat built to the Q, Six-Metre or other class does most of her racing without time allowance and against boats built to her own rating.

This, however, is not true of ocean and long-distance racing. Those branches of the sport were developed and are participated in by boats built fundamentally for cruising, and are scarcely any two of them alike. The functions of the Cruising Club Rule are first to equalize as best it can the chances of a group of boats, diversified as to size, shape and rig; and, second, by various factors and requirements in the rule, to encourage the type of boat most suitable for strenuous offshore racing and penalize the extreme type of racing-machine which is not safe nor suitable for deep water work where bad weather may be expected.

So far this article has avoided one-design classes, but since one-design classes make up by far the larger part of most racing fleets they can hardly be ignored. A one-design class is any group of two or more boats which are identical in shape, rig, weight, equipment and in fact practically every feature that those who made the rule for any particular class could think up. Beyond that, they may be anything. A one-design boat may be a twelve-foot, flat-bottomed skiff used as a training class for children around a club; it may be, like the New York Yacht Club thirty-two footers, a husky, sea-going auxiliary suitable to take a party of five or six offshore; or it may be an extremely fast and pretentious yacht like the old fifty-foot waterline class.

Nor do the names by which most of these classes are known give much of a hint as to their character. A few one-design classes are known by their waterline length or approximate rating designations, like the Thirty-footers, Forties, Ten-Metres and S Class. But with several hundred yacht clubs in the country and with nearly every club sponsoring from one to three or four classes, naming one-design classes is a good deal like naming Pullman cars. There is nothing in the name of the Victory and Inter-club classes to indicate that they are sloops of about thirty feet over all, with small cabins, nor would one assume from the name that Brutal Beasts are ten-foot cat-rigged skiffs in which the kids of Marblehead learn to sail.

Indeed, in the cases of some small stock boats like the Herreshoff 12½-footers, the Cape Cod baby knockabouts, and others, they are frequently known by different names in different ports. The Rainbow Class of one harbor may be the Bullseye Class somewhere else and the Arrows in a third locality, all identical boats. Other classes, like the Stars, Snipes and Wee Scots, are so known wherever they sail, sometimes because of widely spread class organizations which hold national or international championship races.

Lest it appear from the above that the whole thing is pretty simple after all, it might be well to mention that there are several other rating rules, such as the "Square-Metre" and "suicide class" sail area rules. Also that some so-called one-design classes allow considerable latitude in some particulars. Also that no class designation is copyrighted. For instance, the symbol J, sacred to America's Cup sloops in the big league, is also the symbol on the sails of a class of one-design eighteen-footers and doubtless a dozen other small local classes here and there.

In other words it's considerable of a subject. Until some one publishes a 1,000-page book, with illustrations, describing all the classes that exist, with a new supplement each year as new classes are added, the casual follower of yachting will have to be content with what identification he can get from pictures and such descriptions as may be published from time to time of the particular boats that are in the news at the moment.

The table on page 98 is a feeble effort to give the principal dimensions of some typical yachts in the better known racing classes, and of a few of the active one-design classes of the eastern seaboard.

⚓

14

Deep Seamanship

RICHARD MAURY

THE seagoing ketch is two days offshore on her maiden voyage. So far the turn of weather has been favorable. By now the craft is well beyond soundings, a couple of hundred miles out on restless sea. So far the wind has been gentle, and, except for a short heeling squall, flawless throughout. So far the seas, dark blue and sliding by the cruiser, have been moderate and well-behaved. The horizons have been clear. Each night the star-work of the sky has gleamed over the spars from watch to watch. The sea has been paying dividends, and the new crew, with sheets eased, roll happily down wind, their cruiser sliding like a dream over picture-book seas: flyingfish weather, photographer's weather—anything you wish to call it.

Now the lazy wind drifts west. It takes an hour to do so. But that hour over, it comes on stiff. White islands of cloud turn suddenly dingy, and above them is thrown up a grey glare, reddish near the sun, turning to brown near the horizons. The sparkling bouncing water about the ketch loses its glitter. The new crew feels the world changing: an abruptness to the seas, darkness to the north, to the east, to the south, to the winded west. The horizons become noosed by rain. The individual clouds are gone and the grey ceiling expands to every sky line. Something is up! Peace is over. War has been declared, and somewhere out there that old two-faced god, Davy Jones, is looking on with a sly smile.

The uninitiated crew watch a falling barometer, as a pair of flyers might watch their sinking altimeter above a mountain pass. In comes the rain, the wind. The waves are suddenly rolling grey, rising high and coming onto the ketch in packs, row on row of them smoking from windward. The ketch strains, staggers. Here's a real blow already! Shorten sail!

But how much sail should these men take in? And what sails? And with the gale rising what should they do with their craft? Heave her to, let her run for it, or carry on her course shortened down? Run for land? Heave-to on the starboard or port tack? Lay out a sea anchor or sprinkle the ocean with storm oil? Oh, I can tell you, it's a very interesting sight to see two men, or a dozen men, trying to get a new craft through her first blow.

In this instance the crew didn't do much thinking until the main was stowed and the craft was riding it out under jib and jigger. Then they decided to heave her to—the work of a minute! Just lash the helm alee and call it quits! Because it took the best part of a watch to do so, they thought themselves very green and lubberly. In the cruising yarns they had read it sounded ridiculously simple, and while it often is after a long acquaintance with a craft, they didn't know how fortunate they really were. They had tested her in protected water in the eye of an inland wind, and because she hove-to in that instance, they thought she would do it in any weather. They might easily have been wrong. A craft that will lay up to it in thirty-mile gusts might not do so when the wind doubles itself. Particularly in high, broken water. And the greater the amount of broken water the less successful becomes that maneuver. Just as you don't know whether you'll like the sea until you're out on it, you don't know whether your craft will like heaving-to until she's in a real duster. Maybe she'd rather run before it. Perhaps her spars are not stepped to help her, but more likely if she refuses to heave to the trouble is not aloft, but below. The rudder too big or too small. The keel insufficiently deep. Too much bearing aft. Most likely of all a lack of forefoot (some of these cutaway, fin keel craft appear to be in mortal agony when laid up to a bit of wind). And if a craft heaves-to on one voyage and refuses to on another, you can wager the trouble is neither alow nor aloft, but somewhere between keel and mast. She's carrying her load badly. Too many stores aft. She's trimmed by the stern, and every sea throws her bows down as she pivots on her over-weighted quarters.

A few broad generalities: it's almost impossible to heave to a truly fin-keeled type of boat. The greater the water line length in proportion to the over all the better. The more buoyant the craft the less inclined is she to heave to. The greater the displacement forward the easier becomes heaving to; the less the forefoot the harder. (But the deeper the forefoot the more dangerous is it to fly under bare poles.) Broadly speaking, a slow craft heaves to easier than a fast one, and, just as broadly, is less safe running before it.

That brings up an interesting question. Which is the greater virtue in a ship—to be able to lie up easily or to run safely under bare poles?

But before attempting an answer, a few more words about heaving to a new and untried craft. In a two-masted cruiser, the new crew will take in all sail but a small amount forward; then they will put the helm alee.

The rudder will tend to bring the bows into the sea. But the rudder will do so only for as long as enough sail is set to move the craft through water. Should she stop, should the sail be insufficient to hammer the craft forward, then the crew may just as well take in the rudder also. A cap of wind hits the sail (in a small ketch it will be a reefed main or perhaps a big jib and a jigger, in a large one a storm trysail, in a schooner a close-reefed foresail, in a yawl or a sloop a trysail or a snug jib). The craft goes off the wind (she cannot safely go off more than six points, and, idealistically speaking, should not yaw five points) and as she gathers more and more headway, the rudder, becoming increasingly effective, brings her up again; brings her up until the sail begins to spill wind. As it does, steerageway ceases, the rudder loses its effect and the sail once more drives the bows off-wind. The process is repeated.

It is a mild form of punishment, but it will be seen that the one flaw in the maneuver is that in a heavy gale enough canvas must be set to insure steerageway. Yet there must not be too much up there, lest it out-master the rudder and sheer the craft broadside to the blow. The problem is to effect a good balance between sail and rudder, a problem that varies with the individual virtues and defects of the craft.

Remember some of these hints. You may not have to lash your helm hard alee. Some boats lay up to it with remarkably little lee helm. If the sail overpowers the helm do not try to compensate for this by giving too much sheet. The sail will be flogged badly as it comes into the wind, the leech will suffer, and the craft is likely to come into irons, perhaps to pay off on the other tack—a dangerous thing, as the helm will then be hard aweather. If the bows are driven off-wind by heavy seas, as often happens in small, light craft, try stowing some heavy gear in the forepeak.

Very often this feat of balancing will need the aid of an additional piece of cloth, either amidships or further aft. This helps the rudder to pinch into the blow. You're not likely to need anything as large as your smallest jib for this. (We were once helped by setting a blanket in the main rigging of a schooner, and I've heard of a 1,500-ton sailing ship that would only heave to when her weather dodger was rigged to the railing of her poop; and of another great ship that was hove to off the Horn by setting only a strip of canvas in her lee mizzen rigging.) If this balancing cloth is set off the mast, it should be on the weather side, and if on the mast, sheeted home snugly, or even slightly to windward. If all this fails don't despair. Wear around and try laying up stern first.

Heave to on the tack that carries you farthest from shore. If you expect the wind to back suddenly, lay up on a port tack; if to veer, on a starboard tack. Finally, when you've succeeded in heaving to for the first time, don't follow the pattern as set down in deep-sea cruising accounts, by immediately repairing to the cabin. It is not the conventional thing on one's first bad night. Watch, watch your craft. Study her. As she falls away, as she creeps into the wind, as she pauses undecided, then yaws off once more. Perhaps for half an hour she rides smoothly over tall, rangy sea water, like a Cape Horn pigeon, resting, riding in that hard wind. Then, suddenly, unexpectedly, she may break your heart by falling off, and, as though she had a hook out to leeward, her bows may sheer around, until the sea is attacking her rail. You watch her—six points off! That's far enough! Here she comes up. No, she's still lagging. *Seven* points off! The gale drives a broadside into the sail. She heels. She makes two, three knots, with a wild, heavy ocean bearing down on her. You go for the tiller, only to remember that the helm is hard alee as it is. Finally it exerts its influence, and up comes the craft, and, with the speed she has gathered in her runaway, she doesn't pause delicately a point or two off the eye of the wind, but rushes courageously into its eye, shakes her sail violently, plunging in irons.

What caused her to yaw? Perhaps just an unusually large sea, a lonely wave, out of proportion to the rest of the pack. Nothing can be done about that. But perhaps it was something else. The crew didn't feel the new edge on the wind that was gradually overpowering the sail, which, in turn, slowly overpowered the helm. Shorten down again. Perhaps the wind shifted a point or two. If it swings far enough off the sea, the crew will have to consider a sea anchor. Perhaps the sea is breaking whiter, hitting the bows harder, to drive them off. They should get an oil bag under the weather bow as quickly as possible. Perhaps the yawing meant that it's coming on too stiff for heaving to, that the time has come at last when it's down sail, and stand before it under bare poles.

To get back to the new crew aboard the ketch. In a month from now, or even less, if they are keen, they'll know quite a lot about the ways of their craft. But right now, good boatmen that they are, they merely attempt to do their best with that unknown quantity, a new boat. With the wind in the forties, the craft lies up to it comfortably under jib and jigger. But the barometer falls yet more. The wind comes on at fifty-odd, a strong gale attaining a pressure of some twelve pounds to the square foot. The ketch no longer behaves perfectly. Her speed is greater. Occasionally she comes nearer the wind, but also she veers much farther off. For one thing, that big mainmast, stepped towards the bows, presents too much windage. She can no longer hold up her head. To ease her, to keep up the bow, the crew take a reef in the jib. She rides easier. But she begins storming dangerously close into the eye of the blow. The canvas shakes. It is obvious that the jigger, together with the rudder, is over-balancing that reefed jib. So, in comes the jigger. The weather increases until a whole gale is booming down from windward. The decks are as wet as the ocean; the light is dying from the sky, and the crew, faced with a heavy night ahead, decide that the loss of the jigger hasn't helped matters at all. Perhaps when they reach port they'll work a set of reef points into the jigger, perhaps a little later they'll try running up a storm trysail. But right now they are going to work fast and set out a sea anchor. They'll heave to no longer.

They had the good judgment not to invest in one of those conical contraptions of light duck worked around a hoop of galvanized iron and sold indiscriminately at most ship chandlers. The hoops are too frail for the job, the canvas too flimsy. Cloth for a sea anchor should be heavier than any sail you set, even heavier than storm canvas. It can be No. oo duck, and *should* be at least of No. 1 stuff. It should not have a seam that is not roped. The best type is not conical at all, but square-mouthed, braced by a pair of timbers. Four eyes are worked in each corner of the roping about the forward mouth. When the two timbers have been set up to brace the mouth, a four-way bridle is secured to the eyes. To this is made fast the cable, which for a nominal cruiser should be of no less than 1½ inch manila.

It is impossible to state the exact size that a sea anchor should be.

Ships don't take to the best of them naturally, and it is a question of testing and more than a little experiment. But, roughly speaking, a cruiser of ten tons should do with one along these measurements: length, 4 feet; mouth, 7 feet square; after-mouth, 2 feet square.

It is a good idea in really heavy weather (and a sea anchor is not of much account in anything else) to secure a patent oil tin, or better yet, an oil bag, to the anchor. The oil will do wonders out there to windward, and cause the craft to ride in a lane of comparatively fair water.

Pay anchor and line over the weather bow. Not too quickly, for if you do, the anchor will drop right under your keel and a long while will pass before it's out to windward taking a strain. As you pay out, stop every now and then and give a short, smart haul. This tends to lift it surfacewards, where it needs to be. Don't expect a mere eight fathom of cable to serve. The anchor won't take purchase until well away from the drifting craft—say at the bitter end of twenty-five fathoms of line—and large, heavy craft may well require 200 feet of scope. But remember that, on the other hand, the weight of too much cable can sink an anchor so far as to destroy its efficiency.

The crew of the ketch now see two good reasons why the cross-timbered anchor is preferable to the conical type. Taking great strains, it is requiring the bracing of the timbers; and, secondly, the timbers tend to keep it from sinking rapidly. An iron hoop would do the reverse. The object is to get it far out to windward, and yet hold it somewhere near the surface, so that it takes horizontal rather than vertical strains. The farther it rides from the surface, the less effective it becomes. If there is a small can buoy on board, the crew will do well to lash it to the anchor. Even a five-gallon can, sealed with paraffin, would help buoy it. However, you don't want it to have the buoyancy to begin drifting to leeward—only enough to compensate for the downward pull of a heavy, wet cable. *Do not use the buoy unless there is plenty of room to leeward.* With it rigged, the drift will inevitably be greater. Its main function is to ease the ship from severe up-and-down strains, and even while making a little drift, to take the vertical strains in the best way.

On a line used exclusively for the purpose, it would help the anchor a lot if several small buoys of cork were secured along it.

One must be most careful in tending the cable inboard of the chocks. No craft likes a sea anchor and will jerk savagely at it morning and night. The strain is so great that the cable should never be made fast to the bitts, but secured about the mast itself. And as the cable will eat through the best of chafing gear, and as it has a tendency to part close aboard, a chain leader should be made to lead from the mast, over the chocks, and there shackled to an eye at the tail of the cable. Someone else suggested this idea: a heavy steel spring between cable and mast to ease the recoil of the bows, pulling heavily on the anchor.

Fastening a tripping line at the after mouth of the anchor allows you to spill the anchor and drag it back on board with little effort. However, it is not indispensable, and may very well prove to be just another piece of unnecessary gear to foul.

When taking in a sea anchor without a tripping line, don't attempt to heave it in rapidly. Take a series of short strains. This tends to set it drifting home. If this fails show some canvas and work up to it. If you can't set sail and yet wish to trip your drag, slip the ring of a stock or patent anchor over the hawser, lower the anchor overside and let it run down the hawser. It will quickly cause the sea anchor to collapse and sink, bringing it under your keel, where it can be easily hove on board.

Don't expect to be able to heave to and lie to a sea anchor at the same time. The craft will either override the anchor or pull her sea gear to pieces every time she pays off. But very often a bit of sail aft will ease a ship bullied by her drag. Helm amidships and a little sail as far aft as possible, that is the best way to stand up to an anchor.

If you're falling in on a lee shore and have no sea anchor, take an oar, a spinnaker pole or a piece of timber, and bend to it your smallest sail. If you have nothing really small, a boat cover, or, in an emergency, a stout blanket will do. Weight the ends of the cloth with, say, a pig of ballast at either corner (if a jib, weight the clew) and, if no ballast, a few links of chain—enough to keep the cloth more or less up-and-down. Rig a bridle to the pole, bend a long line, and lower away slowly. If you want it working almost immediately, show a bit of sail, wear ship, and as you pay out line run to leeward until the anchor is fully away on your weather.

But if that lee shore lies in sight, forget all about sea anchors. They take time to rig and time to adjust. And, lastly, contrary to a popular belief, they will not stop your progress to leeward entirely. In such a position, try facing the weather with all the sail you can show—enough to counteract the drift, to outpace it. Put oil bags at the bow, man the tiller and punch her into it on her only chance.

The Sailor Takes a Wife

ALFRED STANFORD

MARRIAGE has ended sailing pleasures for good in more cases than I like to think of. These wives "just don't like boats," that's the frequent verdict. "The life is too rough," or "Joe shouts so at me, I can't stand it." Or, "the way Tom spends money on that boat—why, a new evening dress is nothing!" These are some of the plaints heard from the distaff side. They represent no superficial challenge! They end in only three ways—each of which makes trouble. Either Tom, or Joe, or you or I sell the boat and swallow the anchor. Or the male sailor goes sailing alone with his male friends. Or the wife drags along reluctantly, while all hands suffer from her martyrdom and sailing becomes less and less pleasant.

If anything, whatever it may be, holds a life and death power over sailing, it certainly deserves a look in even the briefest examination of sailing. If it weren't for the fatal power bound up in a woman's reaction to sailing, the whole matter would be one for hearty laughter. Since the facts will not let us laugh the matter off, let's try having a look at it.

Almost the first definite idea I came across is that this whole business of women aboard a ship being Jonahs, is no invention of modern yachtsmen. It is a matter of long standing superstition. Back in the days of sailing ships there are records of mutiny and revolt because the skipper brought his wife aboard. Many's the old tar who has shaken his varnished hat and pulled at his side whiskers when confronted with a lady aft. More often than not, he has then grunted, picked up his sea bag and gone down the gangplank. It is one of the oldest superstitions of the sea.

Plenty of men feel the same way today. A woman aboard is "bad news." After much digging, the best reasons that can be discovered are that women mean an infringement of liberty and a curtailment of vocabulary. The chief liberty that is curtailed is probably the use of the rail when in need of relief. This is easily overcome if the pressure is really urgent and the sailor is truly a gentleman. He needs only to go to the lee side of the mainsail.

The only curtailment of vocabulary I have ever noticed, if a situation arose calling for a really fluent bit of cussing, is that there was less breath between the oaths when they did come because of earlier restraint.

From these two traditional objections to women aboard, the indictment against women wanders all the way from a noble, chivalrous sentiment such as: "it's no life for a woman," to a yowl of, "for God's sake, why do we have to have salads aboard a ship! It's impossible to stow that damned wooden bowl." With such arguments from the male side, and the kind that I earlier suggested from the female side, there are the makings of that special kind of argument where neither side can possibly win. Such arguments are complete circles of futility. They are, in fact, just emotional railing, not arguments at all.

Even if the drastic remedy of selling the controversial boat is finally invoked, nothing is gained. No answer is obtained by throwing the arithmetic book away! Nor is any real or enduring progress made by sacrifice. The animal is only anesthetized. It will rise to bite again, probably with redoubled fury. Not only are all the conventional "arguments" on this ancient subject very suspect, but the plain fact is that any sailor knows women who are a living contradiction, women who delight in sailing and who are mighty good at it! For there is no reason why a woman, physically, mentally or emotionally sound in other respects, should not like sailing.

Does she "get sick?" Heavens sake, so do men! I can count nearly a dozen ocean racing veterans who get sick enough to resemble a dead cod, then recover, and enjoy a passage—the rougher the better.

Do women "crack up in an emergency?" Nonsense. Women are wonderful realists, if they are given a chance to be. Who is the practical little gal who runs the house and the children with such a level head? However, expect them to be hysterical, act toward them as if they were going to be hysterical, and their response will be little different than a man's response if he were so treated.

Do women "have no stamina?" "Can't take it?" That too, is twaddle. Physiologically it is known that women are better equipped to stand pain than men. As for endurance, who are the noted distance swimmers?

Do women "gossip and make trouble in the crew"—"talk about personalities at the yacht club?" Holy cow! What mere female chitchat can compare to the intensity of interest with which two hearty male sailors will relish a juicy bit of stray talk about one of their colleagues?

Perhaps it is possible to translate some of the woman's railing against sailing. What she may mean when she says, "I don't like sailing" is, "I don't like to do the scut work aboard while you play god." Other translations can also be made. "I don't like to be the one who always has to remember the race circular and which side to leave the marks, while you take the tiller and steer the boat across the finish line with a look of triumph on your face." If it is a cruising family situation, further translated remarks might run: "You can rant all you please about the cook being the most important job in all the crew, but if it is so, how would you like to have cooking and cleaning up below as the exclusive use of your talents."

Male or not, I for one must reluctantly admit the truth of some of these "translations from the Chinese" of women's railing. I'll be damned if I'd like sailing either, if I had all the housekeeping chores and none of the fun of sailing the ship. What can be done about it? If men get a freedom, a sense of release—in short, fun—from sailing, isn't it wholly reasonable that women be allowed to share in the fun and adventure as well as endure the repetition of their shore work only under more arduous conditions? What can women do, then? Are there any jobs for their "puny arms and delicate hands?" Jobs that really center in the ship's life?

Once granted an open mind and a willingness to experiment, all kinds of ways for the woman to participate in the life of the ship and contribute to it crop up. The first and most obvious department of the ship's life a woman can happily fit into—the heart of the whole business—is the helm. Usually, quite living up to their reputation of being highly intuitive creatures, women will catch on in this very fundamental department of the ship's life with surprising ease. Then the woman immediately has status as part of a deck watch if you are cruising or as an alternate at the stick if out for an afternoon race.

It is probably necessary to make one small suggestion in this respect (and it's almost as true when you're breaking in a man at the helm)—

don't give too much advice! Let the ship and the sails tell your lady helmsman what to do. If the woman happens to be your wife, by all means go below when she takes over in the early stages of learning. The temptation of a bewildering stream of advice can better be resisted this way. Better read a good book too, so you won't be rushing up on deck. Let some old trusted hand smoke his pipe somewhere abaft the wheel, just to answer questions. Sail trim, wind shifts, the problem of not pinching up too high to windward, the painful results of sagging off to leeward—all these are quite a lot to take in along with remembering compass points or racing rules.

Now as we have agreed, no one can be expert at the helm who doesn't understand sail handling, whether it's cruising or racing, small boat or large. So it is the next step in wisdom to throw aside the exceedingly doubtful cloak of chivalry that makes one say—"What! Let a woman go forward? Never!" Stop and think. Even a woman can be useful up there. How about bringing a sorely needed extra stop out to the bowsprit, letting go a sheet, easing a halliard down, finding a handy billy, bringing you a knife or a marlinespike? Being one of the men up forward will help her visualize your crew's problems during a sail change, so that you can safely put her at the wheel some time when you need all hands forward on a really tough job.

The next most obvious job for a woman is that of navigator. Here, surely, bulging muscles and a strong back are not at a premium, nor quickness and dexterity of hand and eye a handicap. Again, too, from her role as navigator, a woman can witness and participate in the whole strategy and life pattern of the ship. She can find the deep excitement of voyaging over endless miles that otherwise look just the same acreage of water, day after day. She can share in the thrill of finding the lee mark first in the fleet if it's just afternoon racing. Her chart work will be neater and more exact than the experienced sailor's because plotting a course is not an old story to her. She will feel her responsibility. The parallel rulers are less apt to slip in transferring a course and the pencil she uses is apt to be well sharpened. She is apt to look twice at the compass rose before calling off a course.

Beyond these suggested avenues of treating women as humans, still other possibilities loom up. I don't mean such obvious other chores as sewing up a torn spinnaker in a hurry or remembering for you where you left just the snap shackle that you want.

The next jobs illustrate that great category of jobs every sailor will admit is of great importance. I mean those jobs which are usually neglected or only half done because "there just wasn't time" for them. There is the field of radio, for example. Beside the weather reports, news broadcasts and time signal checks, a lot can be picked up now by short wave, listening in on the fishing fleet.

Then for a real super-woman there is the field of code and blinker. This rich and generally neglected field brings all the steamer talk within your knowledge. All this suggests better and more frequent use of the radio direction finder, now cheap, compact and portable enough to rate consideration even on a rowboat. Whenever you ask a skipper or navigator: "How do you like your direction finder– Do you have any luck with it? Do you really trust it?" the answer is usually a drawled, "Well–." This inconclusive trust is due to the fact that he really does not use the instrument except in fog. Then there is a minimum chance to check its accuracy. So the distrust and uncertainty continue.

Actually, a direction finder should be regularly used on clear days, within sight of lights and buoys, again and again. From such constant use the voices of the various beacons and their peculiarities become as familiar as the lights and lightships themselves.

In *Vision,* we use Cornfield Light Vessel at the eastern end of Long Island Sound as a homing beam frequently. The special tone of its voice with its low tone and high notes, its synchronized fog horn that has a characteristic grunt at the end, all these are extremely useful because we know them so well. A few others, such as Seal Island, Mt. Desert, Cape Cod, Nantucket, and Lurcher have been etched into our memories by special circumstances. But when it's clear there is "never enough time" to bother to set up the direction finder, take bearings and plot them. Thus a much better ear acquaintance with these potentially highly useful aids must wait until chance or harsh necessity force the acquaintance.

A well-known skipper told me with exasperation that the very next trip he went on, he wanted to ship a man without hands or feet—someone who could not possibly be of use on deck—and station him somewhere with a camera with only one job, just to take pictures. Just how this mythical fellow would be able to hold the camera or snap the shutter would still have to be discovered. Yet this same skipper has a wife. There is "boat trouble" between them. She stays ashore with the children while he sails. So—among other things—why not give the little girl a camera?

Have that scene jibing around a lee mark in the squall forever to look at and show. Recall, on some lovely day in winter, the heat of an August day when your crew were sprawled here and there over the deck and there was barely a ripple on the water. How happy all this sounds, and so obvious, most of it too. Not only are these possible pleasures often missed, when women rate only as scuts aboard, but something even worse happens. There is the infinitely varied role of martyrdom—which even the best natured of women can find congenial. It is just around the corner from the galley stove. It is always lurking in the cabin shadows when a wife is turned into a pot scrubber.

In a boat, the dirty jobs never end, the bad breaks that can depress the spirit of all hands never cease, so the opportunities for martyrdom are perhaps the most unlimited of any conceivable environment. When a sailor has a brave, enduring person aboard who can take hours of rain and cold at the wheel cheerfully, how short-sighted to let her turn into the most unpleasant companion imaginable merely because of a damp bunk and no real part in the life!

Most of the contributions so far suggested, it will be observed, while they place the woman in a dignified, one might almost say human, status aboard are definitely not in conflict with her remaining a woman. The most tiresome woman in the world is the woman who wants to be a man. Given a chance to participate as a woman, it seems to me the female should be willing to give up this ambition that is so directly in conflict with the biological facts. Wanting to be a man leads a woman to talk tough, to want to be one of the boys, to tackle jobs that are beyond her strength, get in the way in emergencies and be a general nuisance if not a danger. As a woman, she can be easily and harmoniously a part of the ship's life; as a would-be man, she is a misfit from the start. This one false orientation can lead her, like most phantasy parts, only from one exaggeration to another even more uncomfortable, unpleasant and unprofitable.

Then too, as a wise woman, the little girl will quickly realize that there are times when males like to be alone. Staying ashore now and then on a particular trip will only emphasize the advantages of the better food, the interesting photographs, the news and time regularly reported, the navigation that is up to the minute, the extra relief at the wheel, and the many messenger jobs on deck that are so very handy when done and so truly missed when neglected.

Probably the most critical feminine decision the woman aboard has to make, and one of the hardest for some of the fair sex to understand sometimes, is that there has to be a boss at times on a boat. Frequently, for better or for worse, for wiser or for less wise, a man must take charge. There are times when there must be definite orders and there is no time or mood for discussion. Backseat driving at sea is not just irritating, it's fatal.

⚓

16

Cutters and Others

L. FRANCIS HERRESHOFF

I DROPPED in at the yacht club one evening last week. It was strangely quiet in the library. But presently I heard the voices of two well-known characters. They must have been sitting quite close to the door for I couldn't help overhearing them. One was Mr. Jovial Conversation and the other, who was a little older, was Mr. Precise Commonsense. I had not heard the beginning of the conversation, but my attention was suddenly taken by these words: "Cutter, be damned!" said Precise. "There hasn't been a true cutter in this harbor for years, and what's more there's not likely to be one again."

"Yes, I know," said Jovial, "but they all call 'em cutters now."

"No doubt," said Precise. "They also call a forestaysail a jumbo, a shroud a stay, etc. . . . only show their ignorance; damn confusing at times too."

"Well," said Jovial, "what *is* your definition of a cutter?"

"A cutter, young man," said Precise, "is a single masted vessel always rigged with three headsails, a running or reefing bowsprit, a housing topmast and a gaff; and, of course, sets of topsail. She generally has a loose footed mainsail and sets her jib and topsail flying, the latter generally on a yard. The cutter was first developed in France for the revenue service. Their first ones were pulling or rowing boats from large naval ships; hence their name. As time went on, and I am speaking of about the year 1700, they developed into a rather fixed type of small vessel supposed to be able to overhaul the cross-channel luggers and other smugglers of the time.

"In order to do this, they adopted a sloop or single masted rig so arranged with a reefing bowsprit and long topmast that an enormous sail

Reprinted by permission of, and copyright 1943, by *The Rudder*.

spread could be made in light airs, and still be capable of shortening down to a very small sail area when caught in heavy weather. Thus we have the cutter which is a specialized rig on a sloop. In other words a cutter *is* a sloop and has always been called so in France, the country of their origin."

"Well," said Jovial, "I don't see any plain distinction or norm between the two."

"Maybe you don't," said Precise, "but in no stretch of the imagination can you call the wide, high sided tubs of today cutters. You take the modern sloop—no bowsprit, no topmast, no gaff, forestay inboard, one or two headsails. They are plain sloops. They have none of the peculiarities of the cutter rig."

"Well, I can't see any harm in calling them cutters if the youngsters get a kick out of it," said Jovial.

"Why, now, I don't know," said Precise. "Take the cutter—it has a distinct place in yachting history and the development of our modern racing yacht. How could we have had the heated cutter-sloop controversy between 1880 and 1890 if the difference were not important? How could we have developed the compromise type: first *Shadow,* then *Puritan,* etc., etc.? But this brings up the matter of hull proportion or model. Some time after 1800 Great Britain went in for taxing her shipping more exactly, so she evolved several tonnage rules, one after the other, and most of these rules left a distinct mark or, shall we say, influenced the model of sailing vessels. One rule which measured the length on the keel started the raking rudder post. Another, which measured the distance between perpendiculars, favored the plumb bow. Sometime around 1830 the so-called Thames Measurement Rule came into use. This rule was roughly the length between perpendiculars, times the beam, times (strange to say) one half the beam. It is said that half the beam for the third factor was adopted because for many years the vessels measured averaged a depth of hold of one-half their beam, and cargo in the hold, etc., made the depth measurement often difficult. Well, as beam was measured twice and depth not at all, naturally the English ships became narrow and deep. Although many do not know it, this rule nearly drove English shipping off the seas by 1850 for her clipper ships were small, narrow, slab sided craft that the large, liberal beamed American clippers generally beat even with a larger cargo aboard.

"But to get back to the cutter yachts, or perhaps we had better say

English cutters, they were rated for racing purposes by this same Thames Measurement Rule. This rule produced, or encouraged, a deep narrow vessel with plumb bow and long overhanging stern, as the stern was not measured aft of the after perpendicular which was at the rudder post. Now, as sail went free, they used the cutter rig which allowed an enormous sail spread with its long reefing bowsprit and a topmast as long as the mainmast. They set a large English topsail, shaped like a standing lug sail, with the yard crossing the topmast. In anything but a calm they sailed heeled well over and in any breeze at all had to shorten down considerably which they certainly could do in endless numbers of combinations, so that you will often see them in old prints with the topmast housed and small jib set half way out on the bowsprit with the tack in the pigtail of the ring traveler, which was adjusted by an outhaul."

Jovial then said, "I have often heard it said that a peculiarity of the cutter was that her mast was farther aft than a sloop's."

"That is definitely not so," replied Precise, "and I have checked it on many sail plans and photos. You see, the thing which locates the center of sail plan and the position of the mast is the underwater profile of the yacht; and the cutter, with her deep sharp forefoot, required the sail plan unusually far forward. While it is true that some of the later cutters, like *Minerva,* were cut away forward or had great rockers to the keel so their lateral resistance was far aft, still as a general rule the cutters' masts were no farther aft of the stem head than on the average sloop. Of course they both had various steppings of the mast and this was partly dependent on the size of the fore triangle carried, be it either sloop or cutter."

"Well, how is it then," said Jovial, "that the naval architects themselves call 'em cutters? Take the So-and-So's Seventeen Seas Flying Cutters, or the What's-It's-Name-Company's Deep Water Cutters."

"Well, sir," said Precise, "they are naval architects—you couldn't expect them to know about yachting history. Great big fellows, you know, rather out of their element working on the designs of yachts. But if you ask a real yacht designer he will tell you soon enough the difference between a cutter and a sloop, and if he is an American he will not be ashamed of the American sloop, which has generally proved itself the fastest sailboat in the world."

"Now how about the marconi cutters, jib headed cutters and Bermudian cutters?" said Jovial.

"Yes," said Precise, "it was a pretty hard pill for the Britishers to swallow when the leg-o'-mutton sail came into popular use. You see, since Colonial times, our common rig was the leg-o'-mutton sail for small craft—the Bugeye, the Connecticut River Sharpie, the Block Island boat, the Isle of Shoals boat, to mention only a few. Some had short gaffs or large headboards and some were thimble headed, but to all intents and purposes were leg-o'-mutton rigged or had but one halliard.

"The popularity of this rig extended to the West Indies and even to Bermuda a hundred years ago so that when the leg-o'-mutton sail came into popularity again the Britishers called it the Bermuda rig or anything but its American name—the leg-o'-mutton sail. As far as the marconi part of it is concerned, that is most amusing. It seems that about 1912 or '13, when the English large sloops or compromise-cutters adopted a pole mast, or one continuous spar for the mainmast, topmast and topsail yard, a system of staying was adopted consisting of several diamond trusses like the early marconi radio masts on shore. Well, now, these yachts—*Shamrock III* and *White Heather,* etc.—of course were gaff rigged and set topsails. After the war when some of the smaller yachts were setting leg-o'-mutton sails they too adopted this diamond or marconi system of staying the mast. But at the present time this system of staying has been almost entirely discarded, so that it is both strange and ridiculous to call our modern leg-o'-mutton rigged yachts after the illustrious Italian who never had any connection with yachting."

"I call all of this confoundedly confusing," said Jovial.

"Yes," said Precise, "I call it damn confusing and it only shows the confusion which comes of using wrong names for things. And when you add cutter to these names—marconi cutter, jib headed cutter or Bermudian cutter—you add ridiculousness to confusion, for one of the principle peculiarities of the cutter rig is its long housing topmast. Now how in thunder are you to hoist a leg-o'-mutton or jib headed sail by the hounds, capiron and yoke of the cutter's masthead?"

"Well," said Jovial, "they say a cutter will lay to better than a sloop. Is there anything in that theory?"

"Yes, certainly," said Precise, "but first you must do plenty, as the modern saying is. But now we had better have an understanding about what laying to is so we can talk so the other fellow knows what is meant. As I understand it, laying to means heading about 45 degrees away from the wind with no way on, either ahead or astern, of course making a dead set to leeward, as the sailors call it, which means slowly moving sideways. A small vessel under these conditions is remarkably steady, comfortable and dry; in fact much more comfortable than laying to an anchor or sea anchor. But don't confuse this sort of laying to with jogging, or fore-reaching the way the Gloucester fishermen shorten down for the night with the forestaysail sheeted to weather to kill their way. We will speak of schooners, if you like, after we have considered the cutter and sloop in laying to.

"Now," said Precise, "I once owned the small cutter *Loon* (she was probably named so for her diving ability). It must have been 35 years ago now, and even at that time the cutter had gone by, so many people considered them quite passé; in fact that's how I could afford to buy her. Well, late one summer we took a run over to Nova Scotia, had a pretty good time too. There were four of us aboard, my old friend State Street Jack, myself, the captain and a hand who acted as cook, cabin boy and crew. Well, to make a long story short, after a pleasant cruise along the coast there, we started back one September evening as we had thought it would take a night and a day to make the crossing from Yarmouth to Cape Ann. But it turned out quite calm so our progress was slow for the first 20 hours or so. The next afternoon, however, we had a good easterly breeze and were making good progress, but the wind was increasing and slowly veering toward the southeast. We took in one sail after the other. By four o'clock it was blowing a strong breeze and we were maybe 30 miles from Cape Ann. The captain and I decided we didn't want to make Cape Ann at night in a southeaster, so we agreed to lay her to. Now this is the point I am getting at. It took us nearly two hours to lay her to. First, while we were still running or wind-on-the-quarter, we housed the topmast. Now housing a topmast at anchor is a simple matter, but in a breeze at sea I'll tell you it is something. The boy went aloft and rove off the topmast halliards, then we slacked up the shrouds and hoisted her till he pulled out the topmast fid. We lowered her away. Then we had a wrestle to seize the topmast heel in place far enough ahead of the mainmast to allow the mast hoops to clear when we got ready to lower the mainsail.

"After lashing the topmast shrouds down we tackled her long noble bowsprit. It had started to spit some rain by that time and the barometer was falling rapidly. Well, it generally is easy enough to house a cutter's bowsprit. Hers had a latch arrangement on the bitts so the heel of the bowsprit could come right aft and swing sideways till the bowsprit's cone or nose just extended beyond the gammon iron on the side of the stem head. Thank God, she had a chain bobstay so we lashed the bite of it up over the bitts. Now the time came to reef the mainsail so we shot her up in the wind and let the whole sail down, which is about the only way you can do in a strong breeze. It was dark by now, raining and real rough. Well, if you have ever tucked a reef in a loose footed sail in a gale of wind you know what we were up against; every time she rolled the whole bunt of the sail would swash down to leeward, taking the reef points out of your hands. No, sir, don't let anyone tell you a loose footed sail is easy to reef in a breeze—might be all right in light weather. Worst of it is, the reefing tackle or clew lines on a loose footed sail take the whole strain and have to be uncommonly carefully done or the whole thing will chafe and give out.

"You may ask, why didn't we set a trysail and be done with it. But a trysail's center of area is too far forward for a cutter to lay to well. All this time we were scudding under bare poles (or running before it, if you like). But now we stood by to round up to hoist the main. I was at the helm, the captain and man forward, and my friend Jack at the cross tackles to cast off the boom the minute she lifted. Quite a sea had made up by this time and as we rounded up and came in the trough of the sea she took a couple of rolls and a good sized comber struck us beam to and boarded us full length.

"Now the weight of water and the wind completely killed our headway so we hardly came in the wind at all, but the men forward managed to get the sail part up, then Jack and I worked our way forward and we finally managed to sway her up between the four of us. And she lay there pretty steady under the triple reefed mainsail, considering the sea that was running, but we were pretty nearly exhausted after all this work. We all went below and had a good drink. Now there are two points I am driving at with all this talk—the first is, yes, a cutter can be made to lay to in a sea if you reduce her wind resistance forward by housing her topmast and taking in her bowsprit. The second is, all these boys who own sloops and call them cutters and think by their use of this name they can make their sloop lay to in a breeze—well it is a lot of tommyrot, and if you don't believe it, go out and try it."

"Yes," said Precise, "the more I think about it the more I realize there are no national rigs or types. Take the English cutter, first developed in France. Take the Norwegian pilot boat. Why, I've known Norwegians to get all excited over them—rather phlegmatic race, too, as a general rule—but you just mention the name Norwegian pilot boat and they will start to breathe hard and their eyes will glisten. You would think someone had said the word 'ski.' To hear them talk you would think their pilot boats combined the speed of a torpedo boat with the seaworthiness of a lightship. Well, what is the Norwegian pilot boat? The best of them, the ones which made the type famous, were designed by Mr. Colin Archer, an Englishman. Seems he had some tubercular trouble, moved to Norway for the drier air there. He was a wonderful designer, no doubt, and won fame by designing Nanson's *Fram;* and, if I remember rightly, the little *Fram* that made the Northwest Passage. Of course Archer had Nordic boats like the Bankfishfartöi to work from.

"Yes," said Precise, "it almost seems sometimes as if the young sailors and some writers love confusion or are carried away with romantic words."

"I have noticed you use the word 'vessel' a good deal in your conversation," said Jovial. "Why don't you use the word 'ship'?"

"Well, to be precise," said Precise, "a ship is a three-masted vessel, square rigged on all three masts, while vessel is a safe enough term to use for anything from the *Queen Elizabeth* down to the well-known thunder-mug."

"I must be going home now," said Jovial, "so good-night."

⚓

17

The Test

ALAIN GERBAULT

The "Firecrest," a little cutter 35' o.a., made a now-famous 101-day passage westward from the Mediterranean to New York. Her one-man crew, the late Alain Gerbault, here tells of the fury of a storm en route.

NEITHER the baffling gales that ripped the sails and set the lockers awash, nor exposure to drenching seas and cutting rains were sufficient to burn the sea-fever out of my veins. A man crossing the ocean alone must expect some distressing times. Sailormen of ancient times who rounded the Capes of Good Hope and Horn, had to fight for their lives and suffered more from cold and exposure. I had a feeling, too, that there was a pretty good chance that some day the *Firecrest* and I would encounter a storm that we would not weather.

The gale continued throughout the night of the 19th of August. Sea after sea swept over the little cutter, and she shook and reeled under them. I was awakened often by the shock of the seas and the heavy listing of the boat.

It was a dirty-looking morning on the 20th, and the climax of all the gales that had gone before. It was the day, too, when the *Firecrest* came near to making the port of missing ships. As far as the eye could see there was nothing but an angry welter of water, overhung with a low-lying canopy of leaden, scurrying clouds, driving before the gale.

By ten o'clock the wind had increased to hurricane force. The seas ran short and viciously. Their curling crests racing before the thrust of the wind seemed to be torn into little whirlpools before they broke into a lather of soapy foam. These great seas bore down on the little cutter as though they were finally bent on her destruction. But she rose to them and fought her way through them in a way that made me want to sing a poem in her praise.

Then, in a moment, I seemed engulfed in disaster. The incident occurred just after noon. The *Firecrest* was sailing full and by, under a bit of her mainsail, and jib. Suddenly I saw, towering on my limited horizon, a huge wave rearing its curling, snowy crest so high that it dwarfed all others I had ever seen. I could hardly believe my eyes. It was a thing of beauty as well as of awe as it came roaring down upon us.

Knowing that if I stayed on deck I would meet death by being washed overboard, I had just time to climb into the rigging, and was about halfway to the masthead when it burst upon the *Firecrest* in fury, burying her from my sight under tons of solid water and a lather of foam. The gallant little boat staggered and reeled under the blow, until I began to wonder anxiously whether she was going to founder or fight her way back to the surface.

Slowly she came out of the smother of it, and the great wave roared away to leeward. I slid down from my perch in the rigging to discover that it had broken off the outboard part of the bowsprit. Held by the jibstay, it lay in a maze of rigging and sail under the lee rail, where every sea used it as a battering ram against the planking, threatening at every blow to stave a hole in the hull.

The mast was also swaying dangerously as the *Firecrest* rolled. Somehow the shrouds had become loose at the masthead. There was now a fair prospect that the cutter would roll the mast out of her, even if the broken bowsprit failed to stave the hole it seemed trying for. The wind cut my face with stinging force, and the deck was, most of the time, awash with breaking seas.

But I was obliged to jump to work to save both boat and life. First I had to get the mainsail off her, and, in trying to do so, found the hurricane held the sail so hard against the lee topping lift that I had to rig a purchase to haul it down with the downhaul; but I finally managed to get it stowed.

It proved a tremendous job to haul the wreckage aboard. The deck was like a slide, and the gale so violent that I had to crouch down in order to keep from being wrenched off the deck and hurled bodily into the sea. I clung desperately to the shrouds at intervals. The broken part of the bowsprit was terrifically heavy, and I had to lash a rope round it while it was tossing about and buffeting the side. Several times it nearly jerked me overboard.

At last I had the jib in, and the bowsprit safely lashed on deck; but it was nearly dusk and I felt worn out. That whipping mast had, how-ever, to be reckoned with, and I could take no rest till at least an attempt had been made to get it tight. So, going aloft on the shaking stick, and clinging to it as it swung from side to side, I speedily discovered that the racking which held the port shrouds in a sort of eye had given way.

Twice I was swung clear of the ship, still clinging to a rope, to be dashed back against the mast with a bang. After nearly losing my hold more than once I found that I was too exhausted to make repairs that night, so slid down to deck to find the whole boat vibrating from the shaking spar.

I feared the deck might soon be opened under the strain of it, so, to steady it, I hoisted a close-reefed trysail, and filled her away on the star-board tack, in order to let the starboard and undamaged shrouds take the strain. I then hauled the clew of the reefed staysail to windward, and hove to.

With this nursing she rode a little easier, and the slatting of the mast was not quite so severe. It was now nearly dark and the gale seemed to be moderating a little, so I went below to get supper.

But when I tried to start a fire, neither of the two Primus stoves would work: so I had to turn in, hungry, cold, drenched and exhausted, for the first time on the cruise sad, fagged out and fed up.

At this point Bermuda lay only three hundred miles south, but New York at least one thousand miles away. I knew, too, that it would be good judgment to head for the islands, and make repairs there before going on to New York; but I had set my heart on making the voyage from Gibraltar to the American coast without touching at any port, and to abandon that plan was heart-breaking.

So much did I feel upon this point that I think I should not have cared if a wave had swept over us and carried the *Firecrest* to the bot-tom of the sea. I tried vainly to sleep. The mast was still slatting about so hard that I feared it would either tear up the deck or carry it away. For some hours I stayed in my bunk thinking the problem out with aching head, and then suddenly decided to try the seemingly impossible.

I got up again, and, as I needed food badly, began to work at the stoves. I filed down three sail needles and broke them, one after another, before I could get one small enough to clear the hole through which the kerosene was fed to the vaporizing burner. And it was nearly dawn before I got the burner working, but I was then able to cook a breakfast of tea and bacon.

This dispatched, I began to feel ashamed of the indecisions of the night before. I felt ready for the battle again, and I determined to sail through thick and thin to New York, the goal of the trip.

On going on deck again I found that, though the gale had moderated a little, it was still blowing hard, and the sea tremendous.

The mast had to be steadied at all costs, and the damaged rigging repaired. It was hard to climb on the swinging stick, and it was harder to stay on it at all. With legs around the crosstrees I had to work head downwards. In that position it took me more than an hour to put a racking seizing round the two shrouds where they came close together at the masthead. Then, dropping down to the deck, I set the shrouds taut with the turnbuckles just above the rail. The mast was now as safe as I could make it.

But there was still the broken bowsprit to repair, and I found it was a job for the carpenter's saw and axe. With these tools I cut a slot in the broken end of the stick, slid it into position and fastened it there with the iron pin that originally held it in place. This gave me a jury bowsprit eight feet shorter than the original one.

As it proved, however, the hardest part of the job had still to be done, for I had to make a bobstay to hold down the end of the bowsprit. This I did by cutting a piece from the anchor-chain, and shackling it into a ringbolt fixed into the cutter's stem just below the water line. To do this I had to hang head downwards from the bowsprit, near the stem, to reach that ringbolt under water. The consequence being that as her bow rose and fell she alternately dipped me two or three feet under water and brought me out dripping and sputtering, to repeat the dose again and again.

I don't quite know how I managed to complete the job, but it certainly had to be done, and, at the expense of many unwilling drinks of sea water, the shackle was got into place and bolted there.

As though in sullen irony, no sooner was this work finished than the gale suddenly moderated. It was just as though the elements were acknowledging that they were defeated, and were surrendering to the gallant little craft.

Taking advantage of this milder weather I made two observations, and located myself in latitude 36.10 north, and longitude 62.06 west. My position was thus about 800 miles from New York as the crow flies, but about 1,000 to 1,200 miles of actual sailing distance.

Although utterly exhausted I was sustained by a keen sense of satisfaction. So much so that I went to work repairing the pump, and soon found the cause of the trouble. A bit of a match had stuck in the valve, and this out I got it working again. After two hours of pumping the boat became clear of water, for I could hear the pump sucking dry; always a joy to a seaman's heart.

Going aloft to make sure the standing rigging was secure, I found the stays had chafed against the mast. It would therefore require careful handling to bring the mast into New York whole. Under the shortened bowsprit and reduced headsails the *Firecrest* was badly balanced, so that when I set the mainsail again I had to reef it by four turns of the boom in order to be able to steer at all; the consequence being that when sailing closed-hauled she made leeway.

I had, however, finished the repair work, so, lashing the tiller, I set my course for New York, and then fell exhausted on my bunk.

⚓

18

The Strength of the Wind at Sea

GARDNER EMMONS

THERE has always been a tendency among yachtsmen to cite unreasonably high figures when describing the velocity of a heavy wind. This arises from the circumstances that a yachtsman's statement of the wind velocity is usually based on guesswork rather than on actual measurements. The fact is that any man who attempts to estimate the *velocity* of the wind simply by exposing himself to it while standing on the deck of a small boat is almost certain to make a bad guess, especially if the wind is strong.

It will help in understanding the question if we consider that there are two distinct ways in which the strength of the wind may be specified. It may be expressed in terms of pressure, or it may be expressed in terms of velocity. Both pressure and velocity are quantities which can be measured directly with instruments. But lacking the proper instruments one is obliged to describe the wind solely with the aid of visual and other sensory observations. The prevailing tendency among yachtsmen, under these circumstances, appears to be to estimate the wind in terms of velocity. That this is a mistake I shall now attempt to demonstrate.

If one is to gauge the speed of the wind correctly, one must be able to determine the distance which the air travels in a given time. Obviously this can be done without instruments (other than a watch) only when the drift of smoke between two fixed reference points separated by a known distance can be observed. These aids are not usually available at sea. It hardly seems necessary to indicate further the absurdity of attempting to estimate the velocity of the wind when the motion of the air is not visible.

On the other hand, the pressure of the wind produces definitely observable effects. The behavior of a boat under a given amount of sail serves as a direct measure of the wind pressure, as does the degree of ruffling of the sea surface. Therefore, the logical way in which to describe the wind at sea is in terms of pressure, or force.

Because of the practical impossibility of observing visually the motion of the air, it is probable that most individuals formulate their velocity estimates from the perceivable effects of the wind pressure. Now, it would be perfectly legitimate to use these effects as a measure of the relative velocity if the absolute velocity were proportional in simple linear fashion to the pressure. Actually, however, the pressure is proportional to the *square* of the velocity, wherefore it is wrong to assume that the velocity has doubled because the "weight" of the wind has increased twofold. Undoubtedly this is the outstanding if not the sole reason why so many people tend to overestimate the velocity in times of strong winds.

To illustrate how this can occur let us suppose an individual gauges the "weight" of a certain breeze by its effect on a specified yacht. He later learns that an anemometer recorded the velocity of the breeze as 14 miles per hour. The pressure exerted by a 14-mile breeze is approximately 0.75 pound per square foot. Subsequently the same individual encounters a breeze which, judging by its effects, has twice the weight of the 14-mile breeze. He then assumes it must be blowing 28 miles per hour. But the velocity of a breeze which exerts a pressure of 1.5 pounds per square foot is only about 20 miles per hour. In other words, the velocity has been overestimated by 40%. Similarly, a wind which has three times the weight of a 14-mile breeze has a velocity of only 24 miles per hour.

The above considerations show clearly, I think, that in the interests of accuracy it is necessary to estimate the wind at sea in terms of its *force*. If desired, the estimate of wind force can later be converted into a velocity figure through the use of appropriate tables of equivalents. Such tables are printed in Bowditch and in most nautical handbooks. Although they are extensively used, it is apparent that very few people have a clear idea as to just how they were derived. For that matter I am not sure that everyone is keenly aware of the fact that the original specifications of the so-called Beaufort scale *made no reference whatsoever to the velocity of the wind*. In fact an impression seems to be prevalent that professional seamen first guess at the velocity of the wind, that they then refer to the table in Bowditch in order to find what number of the Beaufort scale corresponds to the velocity estimate, and that the force number entered in the log is derived in this way. Since, in reality, this procedure is never followed by experienced seamen, it may be interesting and worth while to present a short account of the development of the Beaufort scale and especially the methods employed in determining the velocity equivalents of its various units.

Long before the beginning of the 19th century sailors had evolved a scale in which they specified the strength of the wind through the use of such descriptive terms as "calm," "air," "breeze," "gale," "storm," "hurricane," qualified by such adjectives as "light," "moderate," "fresh" and "strong." Sailors could not define these expressions concretely in terms of velocity, pressure, or force. The scale was based on the effects of the wind on the ocean surface, on the amount of sail that could be carried, on the sound of the wind in the rigging, etc.

In 1806 Admiral Sir Francis Beaufort of the British Navy devised the scale of wind force which still bears his name. The first step in its construction was the selection of a suitable standard or object of reference. Logically enough Admiral Beaufort chose the typical British man-of-war of the time. He then made separate classifications based on the behavior of such a ship and the amount of sail it could carry in winds of different strengths. After long experience and observation he was able to attach to each classification one of the terms then commonly used by sailors to describe the strength of the wind. He noted, for example, that when a sailor said there was a moderate breeze a "well-conditioned man-of-war, with all sail set, and clean full, would go in smooth water from 5–6 knots." The result was a scale composed of 13 units, each of which was described and identified by a term familiar to all sailormen. The original scale, applicable to the full-rigged frigate of the early 19th century, is given in Table 1.

Admiral Beaufort's scale was so carefully devised that a man-of-war could be used to check an estimate of wind force in the same way a standard weight might be used to check the weight of a body. There was no need to know the speed of the wind when such a practical as well as reliable method of estimating its strength was available, and sailors of all nations soon acquired the habit of recording wind observations in terms of the Beaufort scale.

With the development of the various national meteorological services, however, there arose the necessity of specifying the units of the Beaufort scale in terms of velocity, in order that weather reports from ships and from land stations (where it was customary to measure the *speed* of the wind) might be rendered comparable. As a result, independent de-

TABLE 1. THE ORIGINAL BEAUFORT SCALE

BEAUFORT NUMBER	BEAUFORT'S DESCRIP- TION OF THE WIND	BEAUFORT'S CRITERION	
0	Calm		
1	Light Air	Just sufficient to give steerage way	
2	Light Breeze	With which a well con-ditioned man-of-war under all sail, and clean full, would go in smooth water, from —	1 to 2 knots.
3	Gentle Breeze		3 to 4 knots.
4	Moderate Breeze		5 to 6 knots.
5	Fresh Breeze	In which the same ship could just carry close hauled ——	Royals etc.
6	Strong Breeze		Single-reefs and top-gallant sails.
7	Moderate Gale		Double-reefs, jib, etc.
8	Fresh Gale		Triple-reefs, courses, etc.
9	Strong Gale		Close-reefs, and courses.
10	Whole Gale	With which she could only bear close-reefed main topsail and reefed foresail.	
11	Storm	With which she would be reduced to storm staysails.	
12	Hurricane	To which she could show no canvas.	

terminations of velocity equivalents were undertaken by the London Meteorological Office and by the Deutsche Seewarte of Hamburg. The method by which these determinations were carried out was simple enough in principle. Experienced seamen were stationed at exposed coastal or island weather observatories where a full view of the sea made it possible readily to estimate the *force* of the wind. The estimates of these observers were then correlated with simultaneous readings of the *velocity* indicated by the observatories' anemometers.

Now there has always been a strong impression amongst meteorologists as well as laymen that there is a definite wind velocity for each Beaufort number, and that if an observer were actually at the position or level occupied by an anemometer his estimate should agree with the velocity recorded, no matter what the exposure of the anemometer. It was, therefore, somewhat perplexing to discover, when the results of the British and German investigations were published, that there was a pronounced disagreement between them. For example, when the British observers estimated the force of the wind to be 7 the average velocity was found to be 34 knots. But when the German anemometers recorded a 34-knot breeze, the wind force was estimated as 9 by the observers.

It is unlikely that two groups of experienced seamen would consistently differ in their estimates by one or more units of force. Likewise it is impossible that there could have been a national difference of this magnitude, in view of the close contact between sailors of all nations.

A further study of the problem revealed that the discrepancy was due to the difference in the exposures of the anemometers used in the separate investigations. The British equivalents were derived by comparing estimates and anemometer records at five meteorological observatories in different parts of England. The stations selected had unusually free exposures. Typical was the observatory at Scilly, where the site of the anemometer was the highest point on the island, about 130 feet above sea level. The anemometer itself stood about 30 feet above the ground. On the other hand, the anemometers used in the German investigation stood at comparatively low elevations above the ground and the surrounding sea. Furthermore, the exposures of the German stations were not as unobstructed as those of the British observatories.

These facts lead to the conclusion that a good observer does not estimate the strength of the wind by the velocity of the current to which he is exposed. He uses the Beaufort scale to define a state of the atmosphere through a certain vertical range. When a sailor describes the wind as a strong breeze (Force 6) we can measure its velocity and find that on the deck it is 20 knots while at the masthead it is 25 knots. This does not mean, however, that a man aloft will say it is blowing Force 7 when the man on deck calls it Force 6. The estimates will be in agreement, since an experienced seaman gauges the wind strength quite independently of his own situation, and does not change his estimate as he moves from place to place. This simply demonstrates that, whereas an estimate of the force of the wind is not influenced by the point of observation, a measurement of the velocity has little meaning unless the point of observation is specified. In other words, the exposure of the anemometer must be taken into account if it is desired to convert wind velocities recorded by an anemometer into Beaufort numbers.

A method by which allowances can be made for the exposure of the anemometer and, therefore, by which the British and German sets of velocity equivalents may be reconciled is offered by a series of experiments carried out by a German meteorologist. From his experiments it was found that the variation of wind velocity with height above ground over a level grass surface of great extent could be expressed up to an elevation of 100 feet by the formula

$$V = k \ (1.00 + 2.81 \ \log \ [H + 4.75]),$$

in which V = wind velocity in meters per second, k = a constant, and H = height (in meters) above ground. Substitution of various values for H shows, for example, that the velocity at an elevation of 70 feet is about 25% greater than at 20 feet.

It is the set of velocity equivalents obtained in the British investigation that is published in Bowditch and in other American nautical handbooks. Most readers undoubtedly are familiar with them, but for convenient reference they are reproduced in Table 2.

It cannot be emphasized too strongly that these values represent the average of the records of five anemometers, no two of which had exactly similar exposures. By averaging the exposures of these five instruments, in so far as it was possible to do so, it was deduced that the values in Table 2 were those which would be observed at a height of 33 feet *in a perfectly unobstructed exposure over the open sea*. These figures must be regarded as approximate, because not one anemometer used in the tests had an exposure completely comparable to a free air exposure above the surface of the ocean. Furthermore, the sea surface cannot be regarded as a level surface when the wind is strong; therefore the logarithmic law for the variation of wind velocity with height over a level surface will not be strictly applicable.

Of considerable interest are the velocities that would be *measured on the deck* of a ship in winds of different forces. In this connection it seems worth while to mention a series of observations made on board the full-

TABLE 2. VELOCITY EQUIVALENTS OF THE BEAUFORT
SCALE AS DETERMINED IN ENGLAND

BEAUFORT NUMBER	EQUIVALENT LIMITS OF VELOCITY (IN KNOTS)
0	Less than 1
1	1 to 3
2	4 to 6
3	7 to 10
4	11 to 16
5	17 to 21
6	22 to 27
7	28 to 33
8	34 to 40
9	41 to 47
10	48 to 55
11	56 to 65
12	Above 65

rigged ship *Gazelle* during a scientific cruise in the years 1874 to 1876. An anemometer was carried, and regular Beaufort estimates and simultaneous measurements of the velocity were made. As the ship was of the type specified by Admiral Beaufort, the results are particularly pertinent to the whole problem, for the conditions specified by Beaufort could be controlled by the ship itself. There can be no doubt that the estimates of wind force were as nearly perfect as could be made, because the observer could compare directly the amount of sail carried by the ship with Beaufort's original specifications for the various units on his scale. The means of the measured velocities corresponding to the different estimates of force are shown in Table 3.

TABLE 3. VELOCITY EQUIVALENTS OF THE BEAUFORT SCALE AS MEASURED ON THE DECK OF A FULL-RIGGED SHIP

BEAUFORT NUMBER	EQUIVALENTS OF VELOCITY (IN KNOTS)
1	3
2	6
3	9
4	13
5	16
6	19
7	23
8	28
9	34
10	41

It is noticeable that the velocity equivalents for the higher forces definitely fall below the corresponding values in Table 2. This is explained by the nature of the exposure of the anemometer on the *Gazelle*. The anemometer was held in the hand by an observer on the weather side of the ship. It is obvious that the obstructing effect of the masts, sails and rigging on a large square-rigger must have extended some distance to windward, and that even on the weather side of the ship there must have been a considerable reduction in the wind velocity as compared with the air motion at the same level well clear of the vessel. The obstructing effect would be more pronounced in strong winds.

At this juncture the reader may be wondering whether he should use the set of equivalents in Table 2 or those in Table 3 for converting Beaufort estimates into velocities. A brief consideration of the question will clear up any doubt on this score. What the yachtsman is primarily interested in is the speed of the wind in the free air, not the figure that an anemometer held in the hand at deck level would record. Therefore, he should use the equivalents set forth in Table 2, remembering, of course, that they refer to the velocity at an elevation of 30 to 35 feet above the water. If it is desired, for example, to know the velocity at an elevation of only 20 feet above the water, about 10% should be deducted from the values in Table 2. If it is desired to know the velocity at the masthead, an additive correction must be applied to these values, in accordance with the formula for the variation of wind velocity with height. Thus, if the masthead is 70 feet above the water, the values in Table 2 should be increased by about 10%.

It should now be clear that the determination of velocity equivalents of the Beaufort scale, if indeed it is possible to say at all that limits can be rigidly assigned, is by no means a simple problem, for the reason that there is no unique relationship between wind velocity (as recorded by anemometers) and estimates made on the Beaufort scale. But the sailor without an anemometer is concerned only with the availability of a reliable method of estimating the strength of the wind in terms of either velocity or force. As I have pointed out, the logical way is to estimate it in terms of force, *not* velocity. But how can this be done when there are no full-rigged ships to use as "yardsticks"?

With the gradual disappearance of full-rigged ships from the seas it became increasingly apparent that new criteria for the units of Admiral Beaufort's scale must be developed. Much thought was given to this problem in England, with the result that the British Meteorological Office adopted a set of specifications based on the behavior of the typical British fishing smack in winds of different forces. A smack was defined as a "cutter or yawl-rigged average sized sailing trawler, loaded, with clean bottom." Table 4 gives the specifications for the behavior of fishing smacks in winds up to Force 9.

Unfortunately, the criteria given in Table 4 are of little use in American waters. The ideal solution would be a set of specifications based on the amount of sail that could be carried by a standard type of cruising yacht. Since circumstances prevent us from being able to see how a modern yacht would behave under the various conditions specified by Admiral Beaufort for a full-rigged ship, it is not possible by direct comparison to specify the units of the Beaufort scale in terms of the behavior of a yacht. However, an enterprising individual with a scientific turn of mind could, if he were willing to take the time and trouble, make a series of anemometer measurements while under sail at sea, and thereby devise a new scale which would enable one to gauge the *velocity* of the wind directly from the amount of sail carried.

TABLE 4. THE BEAUFORT SCALE SPECIFIED IN TERMS OF THE BEHAVIOR OF FISHING SMACKS

BEAUFORT NUMBER	BEAUFORT'S DESCRIPTION OF THE WIND	CRITERION
0	Calm	
1	Light Air	Sufficient to give good steerage way to fishing smacks with the wind free.
2	Light Breeze	Fishing smacks with top-sails and light canvas full and by make up to 2 knots.
3	Gentle Breeze	Smacks begin to heel slightly under topsails and light canvas; make up to 3 knots full and by.
4	Moderate Breeze	Good working breeze. Smacks heel considerably on the wind under all sails.
5	Fresh Breeze	Smacks shorten sail.
6	Strong Breeze	Smacks double-reef gaff mainsails.
7	Moderate Gale	Smacks remain in harbour and those at sea lie to.
8	Fresh Gale	Smacks take shelter if possible.

With no fishing smacks in sight and lacking an up-to-date scale, in which the average cruising yacht is the object of reference, is there any other criterion available for estimating the wind force? Yes, there is the sea itself. As everyone knows, the ocean surface is extremely sensitive to variations in the strength of the wind and consequently serves quite satisfactorily as a reliable gauge of wind force. In fact, most professional seamen of the present day estimate the Beaufort number of the wind force by the degree of ruffling of the ocean surface, the number and character of the white caps, the amount of foam, etc. They have learned to do this almost subconsciously through experience and through the traditions which have been handed down by sailors of former generations who were in a position to correlate the appearance of the sea with the amount of sail carried by a full-rigged ship. Although it is remarkable how closely professional seamen agree in their estimates, which are made nowadays on the basis of inherited knowledge, instinct and experience, there is a real need for clearly defined specifications based on the appearance of the sea which shall correspond to each Beaufort number. It is, therefore, quite natural that any attempt to draw up specifications of this sort should receive attention; in fact, in 1939, the International Marine Meteorological Commission adopted a set of specifications originally proposed by one Captain Petersen, a German sailing-ship master. Inasmuch as Captain Petersen's criteria may be quite helpful (especially to a novice) in forming a correct estimate of the wind force, they are presented in Table 5.

TABLE 5. THE BEAUFORT SCALE SPECIFIED IN TERMS OF THE
APPEARANCE OF THE SEA

BEAUFORT NUMBER	CRITERION
0	Sea like a mirror.
1	Ripples with the appearance of scales are formed, but without foam crests.
2	Small wavelets, still short but more pronounced. Crests have a glassy appearance and do not break.
3	Large wavelets. Crests begin to break. Foam of glassy appearance. Perhaps scattered whitecaps.
4	Small waves, becoming longer; fairly frequent whitecaps.
5	Moderate waves, taking a more pronounced long form; many whitecaps are formed. (Chance of some spray).
6	Large waves begin to form; the white foam crests are more extensive everywhere. (Probably some spray).
7	Sea heaps up and white foam from breaking waves begins to be blown in streaks along the direction of the wind.

8   Moderately high waves of greater length; edges of crests begin to break into spindrift. The foam is blown in well-marked streaks along the direction of the wind.

9   High waves. Dense streaks of foam along the direction of the wind. Sea begins to "roll." Spray may affect visibility.

10   Very high waves with long overhanging crests. The resulting foam, in great patches, is blown in dense white streaks along the direction of the wind. On the whole, the surface of the sea takes a white appearance. The rolling of the sea becomes heavy and shock-like. Visibility affected.

11   Exceptionally high waves (small and medium-sized ships might for a time be lost to view behind the waves). The sea is completely covered with long white patches of foam lying along the direction of the wind. Everywhere the edges of the wave crests are blown into froth. Visibility affected.

12   The air is filled with foam and spray. Sea completely white with driving spray; visibility very seriously affected.

It would appear that descriptions of the appearance of the sea surface to correspond with each number of the Beaufort scale cannot readily be worded more precisely than this, because circumstances vary considerably. The length of time the wind has been blowing, the area of open ocean over which it has blown, the rate at which it has increased or decreased in force, changes in direction, squalls, etc. affect the appearance of the sea. Depth of water and tidal effects must also be taken into account.

In spite of the somewhat vague way in which they are stated Captain Petersen's criteria appear to be the best available for modern requirements. The most logical way of gauging the wind (for the sailor who has no anemometer) is, therefore, to determine which description in Table 5 best fits the observed state of the sea surface. The Beaufort number opposite this description is then noted. In order to convert the estimate of force into terms of velocity reference should be made to Table 2. That is the procedure which should be followed if accurate estimates are to be obtained.

⚓

Taming the Wild Dinghy

WILLIAM H. TAYLOR

THE dinghy is at once the joy and the terror of all yachtsmen, except owners of yachts so large that their dinghies are dignified by the name of launches. The dinghy is the small craft in which you get from your boat to shore, or vice versa, which frequently spills your landlubber friends into the drink; scars your topside paint; raises blisters and takes off superfluous avoirdupois, and gives you something to worry about when otherwise life might become pleasantly monotonous.

Ever since there were yachts, yachtsmen have been trying to develop the perfect dinghy, and to date the result of these efforts has been negligible. In order to be perfect for certain purposes the dinghy has to be perfectly terrible for others, and vice versa. Some dinghies are better than others, but not much, and not always.

If a dinghy came with your boat when you bought it, you have been spared one of the chief sorrows of the boat owner, that of having to select a dinghy in the face of conflicting advice from every boat-owning friend you have. Still you have the dinghy, which is a sorrow in itself.

Take the matter of stowing a dinghy aboard a yacht. This is a serious problem in yachts up to 60 or 70 feet, or over. If you hoist your dinghy way up in the air and stow it on top of your highest deckhouse, it is out of the way, but its weight, magnified by the height, makes the yacht roll worse than ever. If you set it on the deckhouse forward of the helm, you can't see over or around it. If you set it on the deckhouse it breaks up what would otherwise be lounging space.

If you stow a dinghy on deck anywhere, it is in the way. If you stow it right side up, it soon gets filled with odd gear that has to be shovelled out on deck before you can launch the dink. If you stow it upside down (safer on the whole) it is just that much harder to launch. Whether you stow it right side up or upside down, it is likely to be washed off by a boarding sea unless you have it lashed down so it takes all hands half an hour to get it overboard, in which case it isn't much use for recovering lost articles, passengers, and the ship's cat when they go overboard.

If you carry the dinghy swung out over the side on davits it will sooner or later be smashed or torn loose by a sea, if you go out in that kind of weather. Even if you don't go out in that kind of weather you may have to come home in it. And if you carry your dinghy in davits across the stern you have either to have a dinghy so small that it won't be much use or else it will stick out beyond the sides of the boat and get smashed up the first time you dock.

However, this is comparatively simple compared to the problems of the man who tows his dinghy, his boat being too small to carry it on deck or on davits. If you tow a dinghy big enough to be of any use as such, it will be a serious handicap to your speed. If you tow a dinghy with no cover over it in bad weather it is sure to get full of water sooner or later and break adrift (which is frequently a blessing). If you tow it with a cover on it, you're pretty sure to fall overboard trying to get the cover off when you want to use the boat.

If you tow a dinghy on a short painter, it drags worse than ever and is likely to snap the rope by a sudden jerk. If you tow it on a long painter, it will sheer around and ram any boat that you pass within a couple of fathoms. A towed dinghy is at its worst in a following sea, when, no matter how long or short the line, the darned thing will ride up on a following sea and ram your stern, or maybe even come right aboard (it has happened). Or it will sheer off to one side, heel to the pull of the painter and fill with water.

But no matter how beastly your dinghy may have acted all day, it becomes most affectionate at night when you bring up in some snug harbor for a night's rest. The first evidence of this is when, as you back your motor to set the anchor into the ground, the dinghy painter fouls your wheel and gets all twisted up, requiring someone to go overboard with a knife and cut it free. Thereafter, the dinghy will, at frequent intervals, creep up and kiss the side of your boat with a gentle bump, leaving a mark on your topsides. This is annoying when you're awake and drives you crazy when you try to sleep. A gently-nuzzling dinghy at 2 a.m. of a still night sounds like a drumstick hitting a base drum.

There are cures for some of these troubles, none of which works very well. For a dinghy which sheers around or runs up on the towing yacht, you can trail a long rope over the stern of the dinghy, which makes it tow harder but in more of a straight line, until some passing powerboat crosses astern of you and fouls her wheel in the trailing line.

The affectionate dinghy can be discouraged in a number of ways. Large yachts carry boat booms, but these have to be the length of the dinghy plus the length of a short painter, to keep the stern away from the yacht, and require complicated rigging, and are perilous objects for boats moving around a dark anchorage to foul. You have no idea of the possibilities of the English language until you have heard the remarks of someone who, rowing past your yacht, bangs the back of his head on your boat boom.

Sailing craft used to make the dinghy fast to the end of the bowsprit or of the main boom at night, but in modern boats these spars aren't often long enough to do any good. If you are lying in a tideway, a bucket hung over the stern of the dink on a rope may keep her away from you, until the tide turns or until the wind shifts, or until some harbor pirate steals the bucket.

Many a man, annoyed by the affectionate caresses of his dinghy, has rowed it off to a safe distance, anchored it for the night, and swum back to the yacht. This is a good way to get the dinghy run down, or stolen, but at least you get your night's sleep.

One of the dinghy's big moments is when a landlubber steps aboard. The landsman has two great weaknesses when approaching a dinghy. One is an irresistible urge to step on or near the side of a dinghy, or even on the thwarts near the side. The obvious result is that the dinghy either capsizes or if it is a good dinghy, tips just enough to drop the offending landsman into the water and then bobs up again, as much to say "Maybe that'll learn you."

The other major mistake is to stand with one foot in the dinghy and the other on the float, or with both feet in the dink but with one's weight partly resting by the hands on float or pier. The immediate result of this is that the dinghy slides sideways away from the float or pier, and suddenly the straddler finds himself doing a split to which there is only one possible ending—sooner or later he or she has to let go and drop into the water, climb out and start over again.

Dinghies are like wives. You wouldn't be without them for the world, but they do drive you crazy sometimes.

20

A Long Chance

CARL L. WEAGANT

ONE day in September, 1933, John Alden walked into the office of *Yachting* with a very small photograph of a schooner located way Down East in Newfoundland. He knew little about her but, judging from the photograph, she had character and I was captivated by her at first sight. Her pretty sheer and powerful fisherman lines were the expression of wholesomeness. There was a certain romantic appeal in her graceful clipper bow. Her quarters tumbled home nicely. The broad, cocked-up-like-a-duck's stern would lift easily before a following sea. The ample bulwarks, with heavy frames extending above the deck to a husky rail cap, were ship-like. She was not too fancy—there would be no varnish to keep up, no brass to polish. The short schooner rig, with simple gear, appeared snug, secure. Even her name, *Marit,* was suggestive of simplicity.

Marit was 72 feet over all—large enough to make a comfortable home afloat, that is, if I was ever able to fit her out and go on a treasure hunting expedition, which, I confess, was the idea motivating the desire to have a boat of this size and heft. Drawing only seven feet of water, she could get into Lloyds Harbor and other snug anchorages, and, perhaps of more significance, she was shallow enough to work the West Indian treasure grounds. And her ample beam of 17 feet 6 inches provided plenty of deck space to carry Frostbite dinghies from race to race and also provided space for winches, air compressors, and diving equipment.

With this scanty information and that little photograph, which I perused for hours under a magnifying glass, carried around in my wallet, and sat up in bed at night viewing, *Marit* soon took the form of a dream ship. I deluged her owner with letters of inquiry. The answers added more fuel to the fire of desire. She had been built for Arctic cruising by Captain Victor Campbell, hero of Scott's fatal South Polar expedition and the only man in British history to hold the D.S.O. in both the army and navy. Such a man, I thought, would build a boat right. Apparently he did. She was virtually 15 inches thick.

Built for the ice, birch, because of its hardness, three inches in thickness, was used for planking below the water line. Above water she was of pine. The frames were of juniper, nine inches square, and he said she was "full timbered," a term which I later learned meant that the frames were spaced so close together that they virtually constituted a solid mass of wood from bow to stern. In addition, there was an inside skin about as heavy as the outside planking. All this wood was bound together with heavy galvanized spikes and rivets. She was built to take punishment all right, and before long she took it and I was thankful for every inch of that fortresslike hull.

Convinced that she was perfect and fearful that someone would beat me to her in the spring, I let myself think it was necessary to go to Newfoundland as soon as possible and sail her home. Time was not available to inspect her first and so I decided to take the chance and go to Curling with a crew to sail her home. Too much, I admit, was left to chance, but the idea of an expedition to Newfoundland in quest of an unseen and unknown ship was somehow appealing. And a winter passage would be a new and unusual experience.

The idea appealed to Slade Dale as well and he signed on. Coulton Waugh and Dennis Puleston, shipmates in *Pinta* on her Cape Hatteras cruise, were ready for more. Rufus Smith needed no persuasion. A good crew is half the battle and, with that half of the battle won, plans for the attack on Newfoundland took positive shape by the end of November.

As the train finally descended the divide toward Curling, dusk was settling on the somber black waters which reached in between the mountains like a Norwegian fjord. In the shadows of the mountainous shoreline on the opposite side of the Arm we could make out a schooner. Moored alone in a rocky cove alongside a cluster of fish piers, she was dwarfed by the shaggy mountain which rose behind her. Finally the train creaked to a stop in the hillside fishing hamlet. As we descended to the snowcovered platform with all our duffle, we quickly attracted the attention of an inquisitive group of rustics who turn out for train arrivals.

They assured us that the schooner on the opposite shore was Captain Campbell's yacht. They called her a *yacht* and said she was a fine little vessel, which considerably renewed my faltering enthusiasm.

In haste to view the *Marit* before the light faded, we left our duffle with the station master and scrambled down the snowy slope to a fish pier. There we found a boy with a husky launch to ferry us across. Like a seabird at rest on the water, she rode there with an air of confidence. The spars were clear, bare of varnish. Her planking, while a bit rough, was neatly painted. A white sheer stripe and boot top set off the black topsides nicely. The bow rose gracefully to a well-curved clipper bow and husky bowsprit. Her jaunty stern was broad and beamy. The rigging looked good. So did the sails. These and many more swift impressions erased all doubt.

Eagerly we clambered aboard and began exploring. The mainmast was of Oregon pine, the foremast local spruce. The sails were practically as good as new. There was no rust in the rigging. A rustic but capable barrel windlass was mounted on the spacious forward deck. Most of the running rigging was in fit condition for a sea passage. There was not a single detail about her decks to bring disappointment. The husky gear, heavily constructed trunk cabins and bulwarks lent a feeling of security not found on less ruggedly built boats. She was more of a ship than a yacht. So far she was perfect, but we had not yet investigated for soft spots.

A fisherman, Joe Wheeler, had charge of her for the winter and presently he put off from the nearby shore with the keys. Below, she was pretty well cut up with a number of small cabins and an odd layout. However, I did not expect any more and at least there was plenty of room to try out ideas I had long been nursing about interiors. Of much more importance was the condition of the hull and timbers. We poked and prodded with knives. Crawling into the stern, we investigated thoroughly and found all was well there. Everywhere we found massive knees, heavy deck beams, and heavy timbers—all sound.

She had an engine to boot—a 44 hp., four-cylinder Gray with reduction gear. It was practically brand new and, later, despite the freezing temperature, it started with a few turns of the crank, the starting battery being low. In the store room, aft, there were two excellent compasses and a brand new storm trysail. And she had a toilet which worked.

By this time news of our proposed departure had spread. The prophecies were dire. Advice was plentiful and much of it we took. The dangers of "icing up" were vividly depicted. In addition to overburdening a vessel with its weight, the ice coats the halliards so heavily that they will not run through the blocks and sails cannot be lowered. Pumps freeze. So do fingers and thumbs. And the decks become unduly slippery. We took heed of the warnings.

Advised to be prepared with ice-breaking mallets, Slade went to the mill and came back with four five-foot mallets, their heads made of heavy birch logs. Rock salt was secured to keep the pumps from freezing. I secured a large Quebec heater and half a ton of coal. Installed in the middle of the main cabin, the stove, christened "Red Hot Jenny" by Coulton, made that previously uninhabitable portion of the ship livable. In fact Red Hot Jenny soon became the center and source of life aboard.

By Thanksgiving Day we were ready to sail. One old fellow who was there to see us off expressed the general opinion of the waterfront in these words: "Well, byes, it ain't a question of whether the ship will stand up, but will you stand up?" The local newspaper also stated: "For those without knowledge of this coast a voyage of this nature is a bold venture and it is hoped that the schooner will meet with no mishap on her journey."

The first 40 miles of the passage from Corner Brook, Newfoundland, to New York was in smooth and sheltered water. With the *Marit* shortened down to jib, jumbo, foresail, and storm trysail in place of the mainsail, we headed down the Humber Arm for the Gulf of St Lawrence. Despite the small spread of canvas, a fresh and weighty following wind sent her reeling along through the dark water at good speed. On both sides of the Arm the wooded shores rose to snow-covered mountains. The dull afternoon sun threw a cold, fantastic light on their silvered peaks. Blow-Me-Down Mountain, so named because of the vicious squalls which sweep suddenly down from its heights, stood supreme among the lesser ranges. Overhung by lowering gray clouds, which were racing out of the southeast, the whole winter panorama appeared grim and awesome, yet it possessed an indescribable fascination. We stood on deck in the raw, gusty wind viewing the desolate but majestic scene with exhilaration.

The behavior of the *Marit* was satisfying, too. She steered easily and responsively. Leaving a smooth wake, she forged along as steadily as a big ship. Her easy motion was reassuring. Below decks, Red Hot Jenny made the main cabin warm and cozy. It was Thanksgiving Day and on top of the stove a grand mess of potatoes, corned beef and soup, all stewed together in one big pot, was steaming. Never did a Thanksgiving dinner taste better. When Coulton Waugh played a round of sea chanties on the accordion, the music added the final perfect touch to the feeling of good cheer and festivity.

By nightfall we reached the Gulf and took departure from South Head Light. Several snow flurries passed over, temporarily cutting the visibility to zero and giving us a foretaste of the hellishness of sailing through blinding snow. The fog signal at the light gave out its mournful blasts. There was a heavy swell running outside and *Marit* took the seas comfortably, but not so our too full stomachs. The wind was in the south but showed signs of shifting toward the west. We headed straight offshore to make as much offing as possible before we were headed and before another northwest gale made up. To get as far off that reef-strewn and wreck-strewn west coast of Newfoundland was our main objective. We had been repeatedly warned that in a northwester it was impossible for a sailing vessel to weather the coast.

Before midnight we had made a fair amount of offing, but the wind was shifting fast toward the northwest and blowing half a gale. It grew bitterly cold. Spray began to freeze on deck and in the rigging. The piercing wind bit right through many thicknesses of heavy clothing. Driven snow and sleet stung the helmsman's face. A northwest gale was unquestionably on its way and we were in for a cold beating.

As the wind hauled we were headed farther and farther north toward the Straits of Belle Isle, away from our destination; but it was the only tack that would keep us safely offshore. By this time four of the crew of five were laid low with seasickness. The piercing cold made the devitalization complete. An hour's trick at the wheel drew out interminably. When relieved, we crawled below, feeling half frozen and half dead, and passed out in the nearest bunk or on the floor, where Slade and I kept company. No night at sea was ever blacker, colder or more miserable.

Sometime in the morning we were roused out of our stupor by a call for all hands on deck. As we stumbled groggily toward the companionway, we could hear a sail slatting furiously. The jumbo had pulled out the cleat which held its sheet and the sail was flaying about wildly. Dennis Puleston, the only man who had not succumbed to seasickness, had gone forward on the icy deck and in an effort to grab and hold the sheet had pulled his shoulder out of joint. Bud Smith and I slithered and crawled forward. Ducking the jumbo boom, which was slamming about the forward deck, we managed to lower and stow the torn remnants of the sail. Meanwhile Pills had gone below and Slade was yanking at his arm in an effort to pull it back into place. It refused to return. Sitting on the floor with one foot braced against his shoulder, I made a try at it without success. Finally Coulton did the trick. By that time Pills was as *hors de combat* as the rest of us.

However, the light of morning, gray and bleak as it was, brought a little cheer after the pitch blackness of the night. Revived somewhat by a pot of strong hot tea, we went on deck to wear around on the tack—it was blowing too hard and the seas were too big to go about. On the starboard tack we headed south toward Cabot Straits. Pills had kept close track of the courses and log readings that night and we felt that we had sufficient offing. The wind had settled down in the northwest and it was blowing a gusty gale. Snow squalls passed over periodically. During the night, the forward deck had accumulated a heavy coating of ice and the after deck was covered by several inches of it. We set to work pounding and breaking it off with the birch log mallets which served to crack and loosen the ice from the deck so that it could be shoveled off. In the course of the day we must have knocked off the rigging and cleared off the deck more than a ton of ice.

We made a good run that night, holding west of south to give the coast as wide a berth as possible, and by morning our dead reckoning put us clear and to the south of the coast, which we never saw. Bearing off through Cabot Straits, we headed toward St. Paul's Island. Once or twice we thought we heard a fog signal, but the whistling wind and surging seas made it impossible to hear with certainty or to determine the direction from which the sound came. Kept on edge by fear of running on St. Paul's Island, we peered intently ahead, hoping for a break in the snow. Finally it came, revealing the rugged outline of the island dead ahead. Big seas crashed high against its jagged cliffs. The rising sun occasionally broke through the scudding clouds to cast a wierd crimson light on the bleak hills at the south end of the island while the north end was hidden in a black snow squall. Lonely and desolate, it was a forbidding sight as it loomed out of the snow. We bore well off the southern end and shaped a course toward Cape Scatari.

In the channel between St. Paul's Island and Cape North, the northern extremity of Nova Scotia, the seas were high and confused. *Marit* reeled along before them, bearing well the load of ice which had again, accumulated. The whole starboard side was covered with a solid sheet of ice, about the bowsprit and bobstay icicles of fantastic shape had grown, and the entire deck was white with frozen salt water. Setting to work again with the mallets, we cleared off as much of it as could be safely reached. By nightfall, which comes all too soon in those latitudes at that time of year, enough had been removed to insure safety.

There had been no slackening of the wind that day and with the coming of night it blew harder. Neither was there any relief from the snow. The powerful light on Cape Scatari, from which we had hoped to take our departure, was hidden in the impenetrable white pall. On the basis of our dead reckoning, we gave the Cape a wide berth and then hauled up on the wind to reach down the coast of Nova Scotia, under the lee of which we hoped for easier going and shelter from the unrelenting northwester. However, we quickly found that it was impossible to hold that close to the wind. The spray became so heavy that *Marit* would have iced up dangerously in a short time. Forced to take the seas on the quarter, we headed toward the Gulf Stream. Once in, or near, the Stream, we knew that the warmer air and water would melt the ever-increasing burden of ice.

There was little change in temperature that night, but by noon the next day the ice was beginning to drop off the sides and rigging and that on deck became mushy and was easily shoveled off. The northwester, however, continued unabated and that afternoon we were struck by hard sleet squalls. Both the jib and storm trysail had to be taken in. Under foresail alone, *Marit* was driven faster and faster into the turbulent seas of the Sable Island Bank. The confused seas were unpleasant reminders of the proximity of Sable Island, graveyard of so many fishing vessels. Sailing blindly through the sleet and black squalls and with the wind blowing so hard that fog signals could not be heard, we had only our uncertain dead reckoning upon which to depend. We headed farther out to sea, still keeping an eye and ear out for the island. The tension of constantly watching the black horizon ahead in hope of catching the flash of a light between squalls and of listening intently for a fog signal was finally relieved. By midnight the seas became more regular, telling us that we were in deep water, well clear of the island.

By morning we crossed the hundred fathom curve. The temperature had risen about twenty degrees, and all of the ice had disappeared. The northwester had blown itself out and a light northeasterly sprang up, giving us a fair slant. With the jib and trysail set again, we jogged along easily, no longer worrying about icing up. But by this time we were about five hundred miles east of the coast; if another northwester set in, it might take weeks to make port.

Fortunately, the wind held in the east and south for the next three days and we made up the westing, closing the coast near Cape Sable. Then the wind began to shift toward the west. With the mainsail set and the jumbo, which had taken one day to thaw out so that it could be sewed, and four days to repair, bent in place, we held on the starboard tack into the Gulf of Maine, holding north of west in anticipation of another northwester. But, instead of continuing into the northwest, the wind backed into the south again. Sailing by the wind, we continued on a westerly course toward the coast of Maine.

The sun had not been out long enough to get an accurate sight since we left Curling and we began to wonder exactly where we were. We had been keeping on eye out for a steamer or fishing vessel that could give us a fix, but they had a habit of appearing only at night when it was impossible to speak them. Finally, on the afternoon of the eighth day out of Curling, we sighted a schooner at anchor and headed for her. Speculating on our whereabouts, we each estimated the position and put a cross on the chart. The man who came closest would be rewarded by relief from his watch that night. The five crosses were spread over some forty miles.

Then, as we approached the schooner, a trim Coast Guard cutter appeared, heading directly for us. From her we could get an exact fix. However, we were saved the embarrassment, for directly ahead a buoy was sighted. That far at sea it could be only one of two places, Cashes Ledge or Cultivator Shoal. We were divided as to which it was. Heading straight for the buoy, as if making a perfect landfall, we ignored the cutter until we drew close enough to make out the letters CL, Cashes Ledge. Then we rounded up smartly and waited for the cutter to come alongside, as we wished to be reported. The captain asked if I wanted to come aboard and use the radio telephone, but we had prepared a message in a tin can which was tossed aboard without delay.

With the position well established, we found that Bud Smith had come within five miles of it, taking the prize, and that the dead reckoning was only thirty miles in error—not bad after a run of nine hundred miles. From Cashes Ledge we continued, sailing by the wind, toward Portsmouth, unable to lay a course to the Cape Cod Canal. However, the southerly petered out that night and under power we were able to head her south for the Canal. But before morning we were under sail again with a rising northerly wind on the quarter. At eight that morning we were off Boston. Twenty-four hours later we were in City Island, making the run from Boston to New York under sail in better time than I believe it has ever been done.

As the northerly increased the log began to hum, registering between seven and eight, and better than eight before we reached the Canal. Snow was flying as we entered the Canal at noon and got the signal to proceed through at once. There was no delay and by two o'clock we cleared the western entrance. It was blowing a good, hard thirty miles an hour and, with all sail set, *Marit* started down Buzzards Bay like an express steamer. The tide was with us, the water smooth in the lee of the shore, and it was a broad reach.

By five o'clock we were off Hen and Chickens Lightship in a hard snowstorm. The wind had increased, the rail was down, and *Marit* was fairly steaming through the water at ten knots. Two hours and twenty minutes later we were off Point Judith. It was blowing 40 miles an hour and even harder in the puffs, and she had all she could stand. The wind had settled in the northwest and was blowing colder by the minute. Spray was flying the full length of the boat, icing her up from bowsprit to taffrail. But we were too near home to be concerned and too engrossed in the thrill of driving at ten knots, hour after hour, to think about lowering. We were having the sail of a lifetime.

On through The Race the *Marit* swept in the driving snow and flying spray. Ice gathered higher and higher up the masts, in the rigging, and on the sails. Even the helmsman was iced up, but no one wanted to leave the wheel. Steering at that speed, the cold spray stinging your face and the snow gathering on your eyebrows and freezing on your eyelashes, was unbelievably inspiring. As we neared Cornfield Lightship, the snow ceased to fall and the sky became crystal blue, but a strange haze hung over the water. The lightship, which we passed close by, appeared to be enveloped in white smoke. It was too cold for fog. We did not know what it was. None of us had ever seen or heard of "Arctic smoke," a phenomenon caused by the condensation and freezing of moisture in the warmer air close to the water. As the *Marit* streaked by the lightship, veiled in the smoke and covered with ice, she must have looked to the men on watch like a frozen Flying Dutchman.

We were off Eatons Neck as the sun rose brilliantly over the hills of Long Island. In the clear morning light the *Marit,* arrayed in a mantle of sparkling ice, was a spectacular and beautiful sight. A string of two-foot icicles dangled like gleaming lamp crystals from the headsails. The booms shone like white candy sticks on a Christmas tree. Crusted in glistening ice, the binnacle took the form of a futuristic statue. It seemed a pity to destroy the icy decoration, but it had to come off in order to lower the sails.

By the time the ice had been smashed off the sails and hammered off the halliards, we had reached City Island. The log was taken in, recording a total run of 1207 miles in 9 days and 20 hours—excellent time. The anchor went over at 8:40, just 24 hours after passing Boston— a remarkable run. And we had behind us an experience that will never be forgotten.

⚓

21

Return from the Cruise

H. I. PHILLIPS

Oh, GIVE me a bed that doesn't roll
And a pillow good and dry—
A mattress not like a load of coal,
And far from a seagull's cry!
Oh, give me the feeling of slats again—
(They never seemed much before);
And springs that have never known mist or rain,
Or surf on a shallow shore.

II.

Oh, let me leap into a bed once more—
A bed with no starboard list . . .
A bed that knows nothing of yachting lore
And never's been through a mist;
A bed that's a bed with both ends and sides
And fashioned for goodly girth—
That's never been subject to wind or tides
And never's been called a berth!

III.

A bed that stands far from a cooking smell
And far from a view of docks,
With never my head in a fishing-well
And my feet in a ship's icebox;
Let me hang my clothes in a closet tall;
Oh, I'm soggy and wet and worn

From hanging my duds on a dripping wall
And wringing 'em out each morn.

IV.

Oh, let me undress when I'm on two feet
And not on my hands and knees,
And pull off my shirt with a gesture neat—
And likewise my B.V.D.'s;
I want to get out of a pair of pants
The way any man should do,
And not have to master a midget's stance
Or bash myself black and blue.

V.

Oh, give me a bed with no salt sea's touch—
No breeze from the clammy flats—
A watery bunk doesn't suit me much
With cobbles replacing slats;
I'm weary of rolling with every wave
And wash of some ship's wake wide;
Oh, give me a bed that can quite behave—
And nuts to the wind and tide!

L'envol.

Oh, bed, I am singing your praise this night
And paying you homage now.
For you are not narrow and low and tight
And haven't a stern or bow!
I'm home from the torture that's called a cruise,
And happy I am to be;
So here's to a bed that is damned good news—
You're welcome to those at sea!

22

Lighthouses in the Sky

GEORGE W. MIXTER

"LIGHTHOUSES in the Sky" suggests both the mystery which seems always to have surrounded the work of the navigator and the simplicity of the principles which have made possible that movement of vessels over all the oceans without which the present civilization of our world could not exist.

Broadly speaking, navigation is the art of directing the course of ships in any waters. The subject is commonly divided into three parts: Piloting, dead reckoning and celestial navigation.

Piloting is the art of conducting vessels along coasts and in and out of bays and harbors, where landmarks and aids to navigation are available. Captain Dutton characterized piloting as requiring the greatest experience and nicest judgment of any form of navigation. Mistaken identity of a light or some apparently slight error in the course or distance made good, may wreck the ship, whereas at sea an error in position may be corrected by later observations.

Dead Reckoning, I am told, was originally called "deduced" reckoning, thence abbreviated D-E-D; or in the vernacular DEAD reckoning. It is an element both of piloting and of navigation at sea. As used in modern navigation, it is the process of finding the approximate position of the ship by plotting the courses and distances from the last well-known position, which may have been determined from navigational marks along the coast, or from observations of the sun, moon or stars. This sounds simple, but for various reasons such positions may be inaccurate. Therefore when piloting, the D.R. positions are corrected frequently from bearings of known landmarks; at sea a correction may be made whenever both the horizon and one or more brighter heavenly bodies are visible.

Celestial Navigation is the art of checking or fixing the position of the ship from observations of so-called navigational bodies which include the sun, the moon, four of the planets, and the principal stars. These are the Lighthouses in the Sky.

Up to the time of the Portuguese exploring expeditions sent out by Prince Henry, known as The Navigator, beginning with the discovery of the Azores in 1419, navigation was conducted in a most uncertain manner. Prince Henry, however, collected the then existing knowledge of nautical affairs and founded an observatory, near Cape St. Vincent, to determine the astronomical data required by navigators. Succeeding centuries have witnessed progressive development of those elements of astronomy, mathematics, geography, chart making and instrument design which concern "pilots," as navigators were called in Prince Henry's time.

Columbus had a compass, presumably with its needle on a pivot rather than floating in a tub, an astrolabe and a cross staff, a fairly good table of the sun's declination, a correction for the altitude of the North Star and probably a globe. He could approximate latitude by the same general methods as now used, but the mystery of finding the longitude at sea remained unsolved for about three hundred years.

From 1600 to 1700, charts constructed on Mercator's projection, as are most of those used today, came into use. Napier's invention of logarithms went far to facilitate the navigator's calculations; and the accuracy of observations at sea was increased by Vernier's device and the use of the tangent screw. Continued efforts to determine longitude were without practical success, although that celebrated problem of finding longitude by the method of lunar distances occupied attention of astronomers and navigators.

The advent of the sextant in 1731 often is described as marking the beginning of modern navigation. Previously, all instruments for measuring angles at sea depended upon the plumb line as does the astrolabe, or required the observer to look in two directions at once, as with a cross-staff. The sextant's most important function is to measure angles of altitude above the sea horizon. Because of its small dimensions, its accuracy, and, above all, the fact that it does not require a stable mounting, a sextant peculiarly is suited for use under conditions existing on shipboard. With the compass and the chronometer it shares the honor of being one of the three instruments that have made modern navigation possible.

The sextant's invention most commonly is ascribed to Hadley, an English country gentleman, assisted by his brother who was a barrister. However, at the same time, an instrument of similar principle was invented and built here in your City of Philadelphia by one Thomas Godfrey, a glazier of repute. Watson's *Annals of Philadelphia* suggest that Hadley pirated Godfrey's invention, but the generally accepted opinion is that each man independently invented the sextant. At about the same time, many minds were competing for the prize of £20,000 offered about 1720 by the British Government "for the discovery of longitude at sea." Many suggestions were submitted, but the problem reduced itself to the determination at sea of Greenwich time, by one of two methods: The Astronomer Royal of England contended for the method of lunar distances, while Harrison and others strove to produce timekeepers which would function accurately aboard ship. In 1775, Harrison's chronometer was finally awarded all of the prize of £20,000, and longitude was "discovered." But, as late as 1794 a captain out of Salem was reprimanded for having bought a chronometer—an impertinent extravagance, since the cost of the chronometer may have equaled the crew's wages for the entire voyage. In fact, there elapsed fifty years, more or less, before the comparison of Greenwich time obtained from chronometers and local time obtained from observations became the accepted basis for determining longitude.

Another date of historical interest to navigators is marked by the first publication of Nathaniel Bowditch's *American Practical Navigator,* at Newburyport in 1802. This special abomination of every novice, periodically revised by the Hydrographic Office, continues today as the epitome of navigation of our United States Navy and Merchant Marine. Although confusing as a textbook, it remains almost a necessity for every navigator because of its infinite information and its tabular data.

Thus, during the last century, with the sextant, the chronometer, *Bowditch* and the *Almanac,* there was developed a system which sometimes is called the *old navigation.* Dead reckoning was computed or reckoned by trigonometry and logarithms, rather than found by plotting courses and distances on a chart. A morning timesight of the sun gave a point of position which was correct only if an assumed dead reckoning latitude was correct. The noonsight gave true latitude but longitude of noon position was correct only if course and distances from morning position were accurately known. Perhaps our skipper took an afternoon time-sight, again to check his longitude. Star sights were well understood but less used than at present. These methods continue to be widely used in the Merchant Marine and, although called the *old navigation,* are susceptible, at least in principle, of as great accuracy as are the newer methods.

Since about 1900 there have come into prominence certain changes, not in the fundamental principles of navigation but in their interpretation and in methods of utilizing old principles. These developments sometimes are called the *new navigation,* which now is the accepted practice of the navies of the world and of most yachtsmen, and is superseding traditional methods of the merchant marine. This so-called new navigation had its genesis almost exactly one hundred years ago, on November 25, 1837, in the discovery by Captain Thomas H. Sumner, an American shipmaster, that a single observation, taken at any time, sufficed for determining a line on the chart somewhere on which the ship must be. This is the line of position or Sumner line which is the very essence of modern practice. I like to call them "Sumner" lines, partly because it is a short name which refers only to lines of position at sea, but principally in honor of old Captain Sumner who thus made the greatest advance in navigation since the chronometer was perfected.

⚓

23

The Perfect Ship

WESTON MARTYR

ONCE upon a time I built a schooner. I— But, 'vast heaving! This lordly manner of describing that rapturous business will not do at all. No, I must take a new departure.

Once upon a time the Fates, in tolerant mood, permitted me to set about the shaping of my heart's desire. The Fates, of course, have made me pay for it since, and am paying still. But what does that matter? Build a boat and taste heaven, say I. For once upon a time I built a schooner—and that was a blissful time. Within my mind there had been slowly taking shape, for years and years, the Perfect Ship. Everyone knows her, I expect—that ship. In the minds of every sailor there float pictures of her, each one perfect, each one different; so, when I tell you what *my* ship was like and of how she came true at last— Well, I can hear you. "Very pretty," you will say. "Very nice, and I like that sheer, but if she were mine I'd—" But I will forgive you, for I know I should be just like that too.

It seems to be the business of the ruler of my destiny to heap oppressions on me. He likes to watch me squirm perhaps, or else maybe he thinks these impositions good for me. Happily though, he sometimes has his little joke. Something like this certainly must have been arranged for me, because there is no power that I know of on earth (including hunger) that could force me miserably to toil for years in a New York broker's office. Yet once I did that very thing.

During those horrible years my vision of the Perfect Ship grew very, very dim: and sometimes I nearly lost it altogether. But never quite. And then by chance, one lucky day, I met a fellow-creature. He was, it turned out, a fellow-countryman too, and he worked in a bank some floors above me. George, it transpired, had been born and bred at Falmouth, so we two had not been friends for long before we began to talk about boats. He, lucky man, had owned in his youth a little yawl, and had sailed about in her alone, exploring on his voyages most of those little harbors that lie on the shores of the West Country.

"Those," said George, "were the days. But since then I've never sailed a boat again. And when I think about it, I know I've been wasting my life." And when I heard this my heart leapt, for I knew then that, in that unfriendly land, at last I had found a kindred spirit. And after that, of course, whenever we met, our talk invariably turned towards the subject which both of us loved. We talked of boats and the sea; so much, indeed, that the sea-fever latent within us broke strongly out again, until at last we felt we should have to get ourselves afloat, somehow or other, or perish.

Our first idea was to buy a little boat, in which the two of us could sleep, and spend our week-ends cruising on the Sound; but we very soon found it was difficult to carry through an enterprise as modest even as this one. Pilgrimages to City Island were made in vain. One or two ancient and most unsuitable craft were offered to us; but prices that almost took our breath away were asked for them—prices assuredly higher than those paid for the vessels when new.

When at last it was clear to us that our attempts to buy a boat were futile, such was the strength of the fever stirring in us that we made up our minds to build. But here again we were checkmated, for builders, deluged with large and fancy orders, refused to trouble themselves with such a small affair as ours; and the one firm that condescended to send us an estimate must have figured, we still think, on supplying the boat with a golden keel.

Thus, smothered by too heavy a load of adverse circumstances, did our project die; and one night George and I decided to bury the thought of our little boat for ever. At any rate, that night as George and I were grieving over the grave of our dead hopes, I was suddenly struck by a brilliant idea—which is a most unheard of thing to happen to me. Great ideas are generally very simple, and my idea was certainly simplicity itself; and this is the manner in which it was unfolded.

"George," said I, "how much money have you got?"

"About $10, I think," said George, beginning to feel in his pockets.

"No, I mean—in all the world," I said.

He thought a little while, and then, said he, "If I sold my bits of shares and things, and cashed in everything I've got, I could raise, I think, about $5,000."

"Good," said I. "You're richer than I thought. The savings of my life, with all I can beg, borrow, and steal, would come to about $4,000. Now George, lend me your ear. As we know, the price of yachts in America is appalling. But there are other countries in the world where a well-built boat can even now be bought fairly cheaply. Suppose we bought one, George, and brought her over here and sold her? I think the profit on the venture would make it fully worth our while."

George is a solid soul. "The freight and duty would kill the thing," said he. "Yes—dead. Otherwise it's a sound notion."

"We could save the freight," said I, "by sailing her over here ourselves. And I'll find out about the duty."

"Means chucking our jobs; and you forget the tariff wall they've built around this country. It's high, my boy; and there isn't a loophole in it."

"I'll inquire about the duty, as I've said. But, as for our jobs—are you in love with yours, George?"

"It stinks in my nostrils."

"Then, why worry about throwing it up? If this scheme is as good as I am beginning to think it is— Why, we needn't stop at *one!* We could build a bigger boat out of the profits on the first, and then— You see, the thing goes on *ad infinitum*. Lord! What a life! Sailing boats across the Atlantic—and making our fortunes out of it. Think of it, George!"

"Yes," said George, unmoved. "I'll think of it; but carefully. And then maybe—we'll see."

George is six feet three and weighs nearly a ton, so it is hard to excite him. But when we found there actually *was* a hole in the tariff wall, and that vessels sailing in on their own bottoms, could get through it—duty free—well, then George began to move. And George in motion was impressive, and he proved himself then to be a man of action, which I decidedly am not.

Mine was the idea, but to George is all the credit due for executing it. He wrote, concerning that vital point of "duty," to Government officials of whose existence I had never even heard. And, what is more, he obtained from these imposing personages written replies confirming the good news about that hole in the wall. Then he wrote to all the boat-builders in the world. At least I think he did, for there were very many letters; and in time the answers to them began to reach us. All business correspondence leaves me, as a rule, extremely cold; but these boatbuilders' letters were all wonderful, and I read them with delight. For they talked of such things as limber strakes, spirketting, plank-sheers, hanging-knees, and wales, not to mention reef-pendant cleats and trestletrees; and one man even mentioned a martingale, a word I have always loved. I said so to George. Said he: "Yes. But I want you to read that letter carefully. It's from a firm in Sheldon, Nova Scotia, and it seems to me they are the very people we've been looking for. Martingales sound fine, I know, but the point about that letter is, it contains an offer to build us a 45 ft. schooner for $6,000. And $6,000 is about half the price of a similar boat in New York! In fact, if you think the construction they propose is satisfactory—well, I'm ready, if you are, to go ahead and tell 'em to build."

I did not understand, when George said this, that the dream of my life was coming true at last. Things do not, somehow, happen quite like that, and my realisation of the truth was therefore slow. There were many more letters to send to Sheldon and many more replies to receive concerning details of construction, materials, fittings, and design before the climax came. So a month passed by, and found us still with no definite agreement with the builders and the contract for our vessel yet unsigned. And then, one day when we felt ourselves foundering under a mass of ever-growing correspondence, George stood up and spoke his mind.

"It's no good all this writing. We're in for this thing now, so let's do it properly. You'll have to resign and go to Sheldon and settle things on the spot. Tell 'em exactly what we want—and when they make up their minds what they'll do it for, sign the contract and get them going ahead with the work. You'll be there then all the time while she's building, so that nothing should go wrong. And when you're ready I'll take a fortnight's holiday, and we'll sail her down here and sell her."

Thus did we ingenuously plan. I parted, indeed, from my downtown broker with a very great joy in my heart; and I bought a ticket for Sheldon. But from this precise point onwards a great many things were to happen to us that had not been arranged. For the building of a ship, like the wooing of a girl, is a chancy and uncertain proceeding. And we were to find it so.

For instance: I sat in the train, waiting to depart, with an eye open for George, who was coming to say good-bye. At the very last moment he appeared, laden with large bags, which he threw at me.

"What's all this, George?" said I. "Something I've forgotten?"

"No," said he. "I couldn't stand any more of it. This is my gear. I'm coming too."

The way to Sheldon from New York is long and hard and wearisome. But when the little steamer which plies across the wind-swept tide-rips of the Bay of Fundy brought us in safety in the end to Sheldon's harbor, then I knew that my idea of Heaven was to be realized at last. The bay, its entrance safely guarded by an islet, fitted most exactly my conception of what a bay should be. Upon the clear green water floated sailing craft of many kinds, and the long, low, shapely hills of some Grand Banks fishing schooners held for a long time my delighted eyes. Caulking mallets somewhere near sounded their musical notes. A blacksmith, leisurely at work ashore, sent tinklings from his anvil, as from some sweetly ringing bell; and the blocks of a schooner, hoisting her sails to air, chirruped to us cheerfully like a flock of joyful little birds twittering in her rigging. Past our noses drifted aromatic scents from pine-woods warming in the sun; perfumes from stacks of new-cut resinous timber; the pleasant smell of Stockholm tar, and all the rest of those variegated sea-savours which float about a place of this delightful kind.

There is only one road in Sheldon. It runs along the shore, and on one side of it stand the little houses, each in its own trim plot of garden land. And over the way, scattered along the water's edge, lie the buildings in which the fascinating work of the place is done. Sheldon's one road is short; but to get from end to end of it takes very long. At any rate, George and I found that it did; and a record of our first attempt to walk along it will show the difficulties with which we had to contend. We landed at a little wharf, intending to find, as soon as we could, those builders with whom we had been corresponding. But against that little wharf a fishing schooner lay, and her men were pitching out of her soft and yellowish slabs of some queer substance that glittered in the sun.

And this, of course, had to be investigated. It was cod, we were told, salted at sea aboard the schooner. At the end of that wharf were men working in a rope-walk. They were making a coir cable, and this was something we simply had to stop and see. I shall merely set down a list of the things that George and I had to pass before we arrived at the end of Sheldon's road, where lay the yard of our boat-builder:—

Sailmaker—sewing by *hand* the seams in a mainsail, 75 ft. in the hoist!

Sawmill—in full blast. (I *should* like to tell you what was going on there.)

Cooper—shaping barrel staves (23 a minute).

Blacksmith—making hanks. (Communication between Smith and helper apparently telepathic.)

Irate old gentleman—hoisting dory out of bedroom window. (Reasons unknown.)

Riggers—stepping a schooner's new mainmast.

Black gentleman—putting eye-splice in a wire bob-stay. (Plough steel. A gory job.)

Timber sled—drawn by long-horned ox. (One m.p.h.)

Spar-maker—shaping mast with adze. (Highly exciting.)

Etc., etc., etc.

Now I hope you understand how difficult it is to hurry along Sheldon's road. But we did, at long last, arrive at our builders' yard; and the first things there to meet the eye were three vessels, in various stages of construction, building in the open air. There was a very large man standing upon a stage beneath one shapely counter, and we stopped to watch him as, with an adze, he delicately sliced transparent shavings from a fashion-piece. To us it seemed most ticklish work, for one stroke inexactly made would mar the look of that ship's stern for ever. But the very large man kept on hewing away, most unconcerned and calm, and when, in the run of his work, he came to a large, tough knot, both George and I involuntarily held our breaths. But "chup" went the adze, coming down this time with some extra power behind it, and where that knot had been we saw a clean sweet cut, most beautifully grained.

Said the very large man, "You'll be the gentlemen, I reckon, who want us to build that little schooner. I was glad when I got your letter telling us you meant to come. For we aren't great hands down here at letter-writing; and I guess you noticed that. But now you can show us just what you want, and we'll show you the kind of work we do." Then he climbed lightly down, and, smiling at us, held out an enormous hand. "I'm Brough," said he, "Tom Brough. And I'm glad to see you."

We sat for hours in the kitchen that night and discussed the building of the boat. I drank deep from a jug of mellow cider brewed by Mrs. Brough herself; and "Now, Mr. Brough," I said, as I smacked my lips, "what about the schooner?"

"We haven't got anything settled as yet, in spite of those letters, and we seem to be working at cross-purposes somehow. It's my own fault, I fear, because I kept changing my mind about all the things we thought we wanted. But, anyhow, this is our latest idea. We want you to build a boat 45 feet over all, 34 feet on the water line, with 11 feet 9 inches beam. Depth in the hold is not quite 7 feet, and the draft is about 6 foot 6. We'd like her of oak all through, but from what you write that's going to be too expensive. So we are sticking to oak for the keel, frames, beams, and top-sides, with the best elm you can get for the garboards and bottom planking. Otherwise the construction is according to the specification we sent, except there are a few changes in scantlings which won't make much difference. We agree with all you said about iron fastenings, but we want bolts through all butts instead of spikes. I want a teak deck, but I can't have it, so you'll have to put down the best-looking lot of yellow pine you can find. There's a 9,000 lb. iron keel to cast, and the rest of the ballast, about 3,000 lb., we want in 50 lb. pigs to fit between the floors. I don't *think* you'll find any very great snags in the construction, but there is one unusual feature about the boat. There's a break in the deck, just forward of the foremast, and from there aft till you come to the cockpit the beams are raised 6 inches higher up on the frames, and they've all got a 6-inch camber.

"I think it will look all right, but, anyway, that's how it's got to be, for it's the only way to get full head-room without fitting cabin-tops or a coach roof, which are abominations, or else giving her so much freeboard that she'll look like a hearse. And I hope you won't find there's anything wrong with this construction, for I'm very proud of it, and it's very strong and gives a lot of additional room below. The only thing is, it doesn't allow any bulwarks amidships where the raised deck comes, so that anything lying around there loose is likely to go overboard. But as this boat is really meant for ocean racing, the absence of bulwarks just there is, I think, an advantage, for if she ships a sea it can run off her again in a second or two. There's nothing to stop it. There's a 6-inch gunwale round the forward and after decks, and don't forget the teak rail, for I simply must have a bit of my favourite wood aboard somewhere even if we have got to cut holes in it for the life-line stanchions."

About this time I took another pull at my jug of cider, for all that talking was dry work, and I had been doing much more of it than I usually do. But neither Tom Brough nor George showed signs of saying anything, and merely sat and smoked and nodded their heads at me— so I prattled on:—

"The boat will be schooner rigged, with a single headsail and a jib-headed mainsail 50 feet in the hoist. We can't afford any hollow spars, so you'll have to find us some nice spruce poles. The foresail and mainsail are both very narrow affairs, so we'll have the booms of Oregon pine, as I like some weight at the foot of a narrow sail. And, anyhow, the grain in a good piece of Oregon is always a pleasure to look at. She'll have wire rigging, of course, but no turn-buckles, for deadeyes and lanyards are good enough for us. The staying of the mainmast was a bit of a problem, but I'll show you the rigging plan to-morrow, and it's not such a difficult job after all. The arrangements below decks are simple enough; and for the ceiling and bulkheads you can use any well-seasoned, clear-grained deal stock you like; but the bulkheads must all be doubled, and boarded diagonally on the after side. The doors, panels, and so on will have to be trimmed with mahogany. Not much of it, you understand, but enough to give things a well-finished look below. If we were going to keep the boat I'd say tar her and stain her a good dark brown inside, for that saves all manner of spit and polish. But this won't do, for we've got to sell her; so we'll finish her off in white, with a gold-leaf streak all along her covering board and fancy scrolls at the ends. We'll enamel her ivory white inside and stick in green silk cushions, leather upholstery, and a silver-plated water-closet, and then some ass is sure to fork out an extra couple of thousand dollars for her on the strength of it. Now, George, you talk; for I've said all I'm going to say."

George gave a big sigh. "The varied emotions of this day have been too much for me," he sighed. "Let's go to bed."

"Father says, please come to McAdam's about the yacht."
The assembly, seated for the most part on those long and slippery

benches peculiar to the sailmakers' craft, greeted us with kindly smiles and that benign gesture of the right hand customary with Popes when blessing the faithful, and with seamen when greeting their friends. Tom Brough was there, of course, with three McAdams; and this is a list of some other members of the council, as far as I can recall them:—

Old gentleman. Representing seven sons. All shipwrights, but not present.

Old gentleman's son. The eighth. A shipwright.

Old gentleman's son's son. Too young to be a shipwright yet.

Mr. Bruce. A blacksmith, if ever there was one! Eight years apprenticed to his craft; four years as striker; 35 years as smith. Admits to "still learning something new about hot irrrron every day."

Mr. Bruce's nephew. A smith of lesser repute. Only 25 years or so at the trade.

Mr. Williams. An ex-deep-sea bo'sun. The master rigger.

William. A Barbadian negro, and bo'sun of the riggers' gang.

Mr. MacAlpin, who owns the saw-mill. A timber wizard, about whom much more must be told.

Old John. A carpenter; whose roughest work is cabinet work, and who made the only sky-light in the world that never was known to leak.

McWorth. The sparmaker, who could earn a fortune as a music-hall turn, doing thrilling and perilous things with an adze.

Captain McPhee. Retired Banks' schooner skipper. There for the fun of the thing, I think, but perhaps with some mysterious financial interest in the business.

Mr. MacPhail. A builder of dories; a maker of oars; and a shipper of apples and potatoes to the West Indies.

"Young" Capt. Jennings. Son of "Old Cap." Jennings, ship-owner with four tern-schooners in the Brazil salt cod trade.

"Now, gentlemen," Mr. Pitt went on, "let's see—where are we? We'll begin at the beginning and get things straight. First of all, then, you gentlemen ask us to build you a schooner yacht; and then Tom Brough tells you his doubts about it. That's so, isn't it? Right. *Then,* you say, 'Build the boat and we'll pay all costs for materials and labor, plus profit, of course.' That's to say, *you* pay all costs. Is that correct?"

"We do," said George, "subject, of course, to certain limitations of cost and profit to be arranged."

"How come dat?" inquired William from Barbadoes. "I do'ant rightly unnerstan'—"

"It's this way, Willum," explained Mr. Pitt, "we all agree the boat's not going to cost more than, say, $6,000, and our profit's got not to be more than, say, 15 per cent. We all agree to that first, d'ye see?"

"No," said William, "I do'ant see 'tall. S'pose that boat cost *seben* thousand dollars 'spite of all what we agree. What den?"

"Now, look here," said Tom, "I'll tell you as plain as I can how I see the thing. And then you can take it or leave it, as *you* like. But *I'll* say, right here, if we don't take on this job we're fools."

"I'm with you, Tom," said Mr. Bruce. "We've got to build her. An' we can—as far as her ironwork goes anyway. I'll stand to *that*. There's a four-ton keel, you say, and that'll have to be molded and cast in Halifax and shipped here. An' that new-fangled contrivance for hoisting the fancy mainsail on—a track I'd call it—that'd better be made there too. But I'll do it myself, if you say so. Light and strong it would have to be. But I could do it. All else we'll make here; and I doubt but there's a vast of fancy fittings to her. Well, I'll make 'em myself. Hand wrought they'll be, and you can't do better than that."

"If you'll do that," said Tom, "I guess the rest of us won't have to worry. What do you say, MacAlpin? Her timber work'll have to be very extra special. Can you get hold of the stock?"

"Can I get hold of the stock? Man, I have it," said MacAlpin. "There's timber in my yard that I've been holding on to for years. There's pieces I've kept back that were a deal too good for the kind of rough work *you* put into a vessel. I fair grudged letting you have it, Tom, to hew up rough and all, and ram great ugly spikes through. There's oak logs and knees and natural crooks I've had in the place so long I've grown to get regular fond of 'em. If ever wood was seasoned that stuff is. Have I fit stock, you say? I have, and it's fit for finer work than ever you'll do, Tom, my boy."

"Well, that's prime," said Tom, who I now began to perceive was an artful fellow. "We'll have McWorth making a hollow spar for her mainmast yet. Could you do it, Mac?"

"I could not," said the maker of spars, "and I would not try. There's no need, with good, sound, straight timber growing in the woods. There's a young spruce I·cut three seasons back, that's been pickling ever since. She's seventy feet about and supple, without a bunch of knots upon her. She's light and healthy too, and I reckon she'll need to be, for fifty foot's a tall stick for a little ship. But don't you fret. That spar'll stand all right if the boat can stand *it*—and if Williams knows how to stay her, that is."

"Stay her fast enough," said Mr. Williams. "Stay a little fish pole to hold up a battleship. It's easy, for you've only to rig stays enough. No, what's worrying me is, you can't go cluttering up a fancy yacht with all manner of heavy wire. It'll have to be done neat, and I'll say we can do it if we put our hands *to* it. William and I'll rig up a model and work it all out neat, strong, and ship-shape. If you'd put channels upon her, Tom, though, I'd be easier in my mind, for eleven foot nine don't give us too much spread for all that reach of spar."

"We'll see what we can do about that," said Tom, "and if the rest of you can promise as good as we've heard to date we'll do nicely. But there's one thing we've got to get a grasp of right from the start. And it's this. This craft's going to be a yacht—not a work boat like we're used to. And the point is, we've got to treat her like eggs all through. I mean it's no good planing and smoothing-off and sand-papering and rubbing down with pumice and that and then to go walking all over the job in nail boots. You see what I mean? We'll have to build this boat like a watch and treat her like a chronometer. For a start then it'll never do to build her in the open, for the weather'd get at her and roughen her up in no time, and undo all our work. So we'll have to run up a shed for her before we even lay her keel; and from then on the work will be careful and slow. *Fine* work, you understand, and it won't do to go knocking your pipes out against her paint and spitting baccy juice all about. No, boys. The things we'll have to do—building a yacht—are going to surprise us some. Question is—can Sheldon do it?"

I think it was Old John, the carpenter, who best expressed the sense of the meeting. It was not until the work upon our boat had actually commenced that we discovered what an artist Old John is at his craft; but that evening we might have guessed it, for, under Tom's insult, Old John's artistic temperament reacted splendidly. "Ah! Ye doubt us, do 'e, young Tom," said he. "Ye doubt if us Sheldon carpenters be fit to work at this fancy boat of yours? Aye, he doubts us! So listen here, me lad, and I'll tell 'e summat. Give me the tools and the time and there's *nothing* I can't make out of a piece of wood. But there is one thing I *couldn't* do, and that's make a bollard even out 'o that ugly great lump of dead wood you calls your head. It's too tough, Tom, that's what it is. Too tough and too ugly. 'Tis your job, not ours, to get the work to do. That's what we made you manager for. And 'cos all you could find for us to build was fishing-vessels don't think them's all we *can* build. Boys, what I says is, let's build this little yacht, and show Tom Brough and them Yankees up along the coast the sort of work we Bluenoses can do if we set our hands to it. Build her, I say—and be hemmed to Tom Brough."

Old John's resolution being carried unanimously, Tom Brough winked largely at George and me, and the chairman, after a somewhat lengthy period of eclipse, got up on his feet again, and at once became very businesslike.

The details of this momentous decision, it is true, were somewhat vague, and it took Tom and George and me most of the next day to disentangle them and give them some sort of shape. There was no shadow of a formal agreement nor was anything at all written down and signed, for, in the pregnant words of Captain McPhee, "if we can't trust you and you can't trust we— Hell, what's the use of talking. We're seamen an' you're gentlemen—and there ain't no lawyers *here!*" The main point was that the incorporate craftsmen of Sheldon promised to build us our yacht, and, while they were about it, to put their best work into her. And I cannot do better than quote the old gentleman with the eight shipwright sons on the question of the quality of the materials to be used. "If we do our best the stuff that goes into her's *got* to be good. For we ain't a-going to waste good work on poor stuff. Now, are we? It ain't sense."

For our part we promised to pay for labor and materials at current market prices, plus a ten per cent commission on the total cost as profit to the builders. And the boat was to be built as speedily as circumstances would permit. "About two to three months I guess she ought to take, don't you, Tom?" said Mr. Pitt. "And if there's any details been forgotten, why, Tom and you can fix it all up in the morning."

Such then was the tenor of an agreement which left both parties to it most completely satisfied. It was a great document. The main points of our agreement were, as nearly as I can remember them, somewhat as follows:—

The builders were to supply all labor and all materials which could be procured locally; and we were to buy anything not obtainable in Sheldon and deliver it to the builders' yard at our own expense. The following is a full list of things Tom asked us to order from Halifax, Boston, and elsewhere:—

A 9,000 lb. cast iron keel.

Screw steering gear. Minus wheel, which old John proposed to make himself.

Track and slides for Bermudian mainsail.

W.C. with necessary pipes and fittings.

Cork insulation for ice box.

Enamel, varnish, and a book or two of gold-leaf.

Galley stove. Cancelled eventually, as Mr. Pitt discovered a beauty stowed away in his loft.

Four deadlights.

Six 4-in. brass ports.

Cabin upholstery and cabin lamps were the only other things Tom did not undertake to produce in Sheldon, and George and I decided to wait until we got the boat to New York before installing these latter items. Tom and Old John were to make a model from my rough drawings, and from this model (modified, perhaps, if we all thought fit) the yacht was to be built. Tom promised to carry out all work as quickly as possible, but quality of workmanship was always to be considered before mere speed.

(*"The perfect ship" was subsequently completed, as the book of the same name tells.—Ed.*)

⚓

24

The Road Home

ALAN VILLIERS

It was nearing the end of July when we had the great storm. It began to pipe up from the SSW on the Saturday, when we had been at sea 23 days, and the glass fell for three days. From noon Saturday to noon Sunday we ran 208 miles in a high sea, which was fairly good going for so small a vessel as the *Joseph Conrad*. Throughout Sunday she ran on well in a high breaking sea with frequent long, hard squalls increasing, with the early coming of the cold night, to a strong gale before which we still ran under the close-reefed fore and main tops'ls and the storm fore topmast stays'l.

This day we sailed over the place where Ronald Walker, aged 21, had been buried from the poop of the ship *Grace Harwar* in 1929, after being killed at his work in the rigging. I flew the Australian ensign at half-mast in his memory. This day too we saw a strange and beautiful white bird which was not an albatross; we had none of us seen a sea bird of this kind before, and did not know what it was.

The night brought sleet and snow and greater wind. About daylight next morning, just before nine, a brief lull tempted me to give her the reefed fores'l, for this was fine fair wind and in the great sea her speed had dropped a little. But the lull was illusory, and I could not keep the new sail set half an hour; it came in again without damage. All hands had long grown expert at the handling of recalcitrant square sail even under the most dangerous conditions. Noon of that day—the third day of the blow—brought frequent violent snow squalls blowing fiercely from the WSW, and the sea was now dangerously high. I took in the close-reefed main tops'l, and goose-winged the fore, continuing to run under this minimum of canvas, with the fore topmast staysail to help the steering. But by six bells in the afternoon the gale was such that I began to think seriously of the vulnerability of the decks, the skylights, with open ship below; the weak, big doors. If a skylight were stove in, it would be bad. We had done what we could to protect them, but they were still weaknesses; and the charthouse doors fitted ill. The steel charthouse, a welded job from Ipswich, had not been very well made and the doors could not be strengthened. I had the boys to think of; I had to get round the Horn. I could take no undue chances. I could not stand having to take a badly damaged ship in anywhere, for repairs. That would be the end of the voyage. I wished, on the other hand, to make all the progress I could while the wind was fair, because of the probability of easterlies afterwards, and because I wanted, naturally enough, to be gone from those cold latitudes as quickly as possible.

But there came a time when it was dangerous to run on. The glass still dropped. It became obvious that, bad as it was, the gale itself had not yet begun, though the wind screamed in the rigging. The fierce rolling had caused the compass to swing violently, and steering was difficult even by day. The feeling of the wind on their faces was the helmsman's best guide; but by night this is poor substitute for a compass. Though she still ran well in the great seas, she was clearing their ever-rising crests with less and less margin. Sea after sea thundered at her, broke in a wild eruption of spray and spume and murderous noise, and thundered by: she was foam-covered to the trucks. The wet dome of the sky sat heavy on the mast-heads, never lifting to give any light beyond a gray, wintry gloom: now night was coming down.

I put out oil, and ran on: but this was dangerous. I thought again of the vulnerability of the decks, of all the lives entrusted to my care: should I still run? But if I hove-to, I should drift in the valley of the sea on my way. The headlong rush would cease; no more would the brave ship run down a degree of longitude every four-hour watch, but I should still drift quietly on, and I could lie there in safety, putting out oil. The drift and the scend of the sea would give me two knots. And if I *broached-to,* if I ran too long; if I ran those gigantic breaking seas on board, there would be no morning. Surely the wind would not further increase, the sea not further rise. She had run in safety this long; why not go on? No time to waste! And yet—the safety of all hands was in my keeping. If I made a mistake, I should have a lot to explain before God. Still she ran; I watched. I watched the sea rise, heard the wind increase, suffered the cutting hail. The sweetness of her underwater body was counting now, and she still avoided the weight of the seas with a great cunning in whose contrivance it seemed that man had no part though she ran drier with some at the helm than for others. Her bluff black bows rode down the hollows and rose upon the crests, and her sweet clean counter dragged no quartering waves to upset the breaking seas and bring them upon us, though she ran at speed.

Now the hail squalls close in, and night is coming down. Heave to or run on? Better lie to the night in the trough of the sea, bowing to the storm, than run on and never come to the morning! The squalls march; the wind shrieks; the sea thunders. The ship reels, staggers, lurches, *flings* herself onwards over those high green hills each of which breaks in a flurry of foam as she passes, into the troughs so deep that the wind goes from the sails and screams the louder in the upper rigging. Good helmsmanship alone can guide her now. I have taken in the clew of the fore tops'l and she runs on under the stays'l alone, a rag of wind-stiffened sea-soaked canvas bellied at the bow. The boiling of the foam flying by reeks of murder.

No, no; I shall not run on this night. It is fair wind; but I cannot go on. If any of those great seas break on me; if for a moment one of the helmsmen so much as fumbles with a spoke; if the ghastly tumult of the maddened sea can come on board—these things in this wind are fatal. Fatal! And I answer for 22 lives.

The mizzen stays'l is set, a small sail of stout canvas, to keep her head up when she comes to; I take the wheel, and wait for what smooth, what chance may come. It is the very last of the daylight. The yards which have been squared are braced full-and-by; there is now no sail on them. Off the deck! Off the deck, now! The sailors crowd in under the focs'l head, the only sheltered place. I wait: even with only these two rags of stays'ls on her she runs eight knots. If she comes up to the sea at that speed and runs into a wall of breaking water, God help us all. The oil bags are out—have been out a long time: they help, but nothing can stop a real breaking Cape Horn sea if a ship swings into it with speed. The safety of lying-to lies in the fact that the ship then yields; she gives; she does not fight for headway. She lies in the trough of the sea with her shoulder to the breaking water, like an albatross asleep with its head beneath its wing, drifting and yielding. In great storms this is the only safe procedure. But I had now, with the ship still running at speed through the dangerously high sea, to bring her into the trough; to stop her way, without overwhelming her; to bring her up without giving the sea a chance to break on her while she still had way. It was a serious and dangerous proceeding. A chance, now! I had waited long. The wind screamed in all the stays and rose to a mad crescendo as the ship rolled to windward: sea gushed from all the washports: the very spume lifted from the sea flew frightened in the air. But everything still held. A lull, now! Now! A vicious hail squall had just passed; there came a smooth in the wake of a giant sea that, passing under us, had flung us high and rolled the ship as if it had been shaking a swimming dog. This was the chance I awaited. Down helm! I forced the wheel down and she came; the wake subsided and the foam was not now streaking past. At the same moment, while we still had way, I saw the great sea rise to wind'ard—the great sea, the murdering one: a greater sea than any which had roared by that day. I had not been able to see it before; it had swept down savagely out of the murk and the chaos. I could not see far. Nothing now, but to fight it out. Break, then, you bastard sea! See if you can smash us! It seemed to me at the wheel that I could feel the ship stiffen for the meeting, summon up all the reserves of her great hull's strength to take this shuddering blow. It would be bad; I knew. She knew.

Implacable and murderous, the thundering sea rushed on, its pleasure glinting in light from the tumult of its driving foam: up, up, it towered, and the little ship rose valiantly to meet it. But she could not do it. She could not do it. I knew, as soon as I saw that sea. No oil, no anything, could stem that insane onrush. She rose, rose, rose! But not enough. I stood at the wheel clinging with all my strength to the spokes; for I knew the breaking sea would come there, too.

The sea came, flinging the oil bags contemptuously back inboard before it, snarling and roaring, high above the weather rail. An instant it hung there, and I could see the glint of evil in the foam-streaked green water. Then it broke on board. It might very well have been the end of us. But the little ship somehow avoided the worst of its force. We had taken only a glancing blow. It filled the decks, and stormed aft; it swept over me at the wheel; it drove over the saloon skylight, over everything. Would she lift again? Would another come, while the weight of the first still held her down? She began very slowly to lift, to free herself through the washports of the sea's dangerous load. Her way was gone now. She shed the water, in a long, long time—or so it seemed—and afterwards no more came. The skylights had held. There was no water down below. We were in the trough now, and so lay in safety that long, wild night.

In the morning we saw that the starboard boat was badly stove in where the sea had lifted it on the chocks, as if the stout manila lashings had been cotton, and had cast it aside. Both light brackets were gone; a door of the for'ard house (fortunately one of slight importance, which gave entrance to the lamp-trimmer's small room) was smashed to splinters, and the sea ran in and out among the lamps; and—worst of all—the fore topmast head was badly sprung from inherent weakness which the storm had brought out. The mast was so badly sprung that the topgallant yard and mast had to come down.

I lay hove-to for 20 hours. It was the worst weather of the whole voyage. The tumult of the implacable sea through that wild night was frightening, even hove-to; she lay rolling heavily, but she rode the seas and shipped no more water. It was noon the next day—the Tuesday—before I dared put up the helm, though the wind all this time had been fair. The conditions had eased a little then, and I ran on under the close-reefed main tops'l and fore topmast stays'l, and still put out oil. In this wild weather the fore topgallant mast was taken down, and preventer stays of iron wire were spliced and set up to the topmast head underneath the place where it was sprung; a new fore topmast stay had also to be set up, of iron wire, double. This was a difficult rigging job to do in those conditions, but all hands worked through until it was done. By that time the weather had moderated considerably, and I was able to set good running sail. The loss of the fore topgallant mast and the weakening of the headstays was a serious blow, upsetting as it did the sailing balance of the ship (particularly for windward working). I had to be more careful now than ever, but we sailed from the scene of the accident—some 2000 miles westward of Cape Horn—to the end of the voyage without that mast.

I hoped now we should have no more such gales, though that was by no means certain. The little ship made no water and continued to run on, buoyant and brave in the great seas. The skylight coverings had unquestionably brought her through that night of storm. Even after that, only two of the bunks were sodden. So she ran on, from there to the

Horn. There was not another great gale, though there were minor storms—trifling things, with the wind only pretending. It often happens in those latitudes in winter that, after a severe storm, the weather is settled, more or less, for a greater or less period, depending upon the ferocity of the storm. The worse the storm, the longer the period of settled conditions afterwards. This is no "law"; but I had noticed it on other voyages. So it proved now. We did not have another really hard blow. We had, instead, a succession of southerly conditions—gloomy at times, with fog, and the clammy moisture weeping from the sails and all the gear; sometimes clear for a time but never bright, and for the most part heavily overcast, with the wind always somewhere between south by east and west by north—not often from the west'ard.

We made good progress. There was not again occasion to consider heaving to. In one week we ran 1,348 miles—from Wednesday, July 29th to Tuesday, August 4th. This was an average speed of eight knots for the seven days—not bad for a ship only 100 feet on the waterline, partially dismasted, pretty old, and long in tropic waters with her iron hull gathering the long grass. For such a ship to storm at eight knots for a week in the Great West Winds, through high seas the whole time, was, I thought, a credit to her designers and her builders. She was not built for speed, though she was built to sail well; the primary consideration was safety. To sail at eight knots for a week in quiet water would not be anything to cheer; but a high breaking sea is a serious handicap. In eleven days we ran 2000 miles. I was careful, all this time. I did not try to drive, but to do my best with the ship as she was, and the conditions as I found them. Throughout these eleven days we showed no kites, though I was accustomed to hold to the main topgallant sail which usually did not get becalmed in the troughs. We ran in the ice line, day and night, with double lookouts and the foghorn going, far south of Fifty-five—southernmost ship in the world, then; last full-rigger to make a Cape Horn rounding—staggering onwards with no lights lit, the rust streaks growing on all the paint; the fore topgallant mast below, and only one good boat—that would be useless here if there were real need of it. Nothing now comes to those waters: we were south of the meat ships' track from New Zealand to England. No outward-bounders now thrash round to Chilean ports for nitrate; and I was glad no ship might loom before our path—no ship, and nothing now but icebergs. We ran on unlit through the long nights, for the light brackets were both gone and the lamp room destroyed; had we shown a sidelight we should have lost it.

So we came to the 8th of August, and on that day were off Cape Horn, and by the evening round, and into the Atlantic. Good ocean! All the reefs and all the storms of the great Pacific were behind. How long the tumult of those waters had saved the South Seas from destruction! It seemed almost as if the great belt of the roaring winds had been sent down there by Nature to make the westward rounding of the Horn difficult, and keep the marauding whites from the Pacific isles. For hundreds of years they had succeeded.

We saw nothing of Cape Horn. The day was gloomy and all overcast, with strong wind, and I had no observations. I had meant, if I could, to have a look at Diego Ramirez, to check the rating of the chronometer; but I could not risk it. We saw nothing but one piece of kelp, a large piece drifting by, and the obvious ground swell of graying soundings by the Ramirez bank—a swell that stayed all day and was then gone as suddenly as it came, with both its coming and its going as noticeable as name boards on a street. We must have passed pretty close to get on the bank; we ran too fast to take soundings, and I would not stop. We were probably within five miles of Diego Ramirez; we could only see one mile, most of the day.

So we came round, and the next day being clear, saw Staten Island and all the wild snow-covered hills of Tierra del Fuego, and the tide-swept break of the Straits of le Maire. The boys were splendid. True, they had heard so much of Cape Horn's bitterness that, now we were round, they were almost disappointed. What, only one hurricane? Only one lost mast, one stove boat?

I looked back now at Staten Island, with the Horn behind, and I hoped to God I should come that way no more in sail. I had hoped that before, perhaps even more fervently on at least two occasions: yet here I was again. And I made no promises and took no oaths, for I knew well that my life, such as it is, is too bound up in these ships for me to make forecasts. If the ships bring me here again I suppose I shall come. But I hope not. It is not the sense in these days too often to sail that grim road.

⚓

"Man Overboard!"

CHESTER BOWLES

LIKE every yachtsman who has ever gone to sea, I had often worried about the possibility of losing a man over the side. I had wondered what I would think, and what I would do, and exactly how the missing shipmate could be brought back on board.

In the middle of the Gulf Stream, one day in late April, 1937, I went over the side myself. Our deck was half awash from an unexpected squall. A rather heavy sea was running. I had on rubber boots and oil skins. To complicate the problem even further, I was unconscious, struck squarely on the chin by a heavy cringle in the clew of the fisherman staysail which we were busy getting on deck. Fortunately, I woke up within thirty seconds, kicked off my boots, swam to a life preserver and a half hour later arrived back on board.

Although I had thought a lot about the "man overboard" problem previous to April, 1937, I can say, without any fear of exaggeration, that I have really concentrated on it since then. I have talked about it, argued about it, and now I am even writing an article about it. Logically any article on this subject should start with the best methods of keeping the various members of the crew on board. Perhaps the most important device is adequate life lines.

Some yachtsmen seem to feel that life lines of any kind are "sissy." A really able-bodied sailor, they argue, should be able to get around the deck without artificial support of any kind. Others, who accept their theoretical desirability, still prefer to get along without them, because they may occasionally interfere with genoa sheets, spinnaker guys, etc. Still others feel that a *single* life line is sufficient on the theory that you are not supposed to get hit on the chin.

In answer to these three arguments, I can only suggest; (a) that a great

Originally published in and copyrighted by *The Sea Chest: The Yachtsman's Digest.*

many supposedly "able-bodied" sailors have already fallen overboard; (b) that it takes only a few seconds to adjust the lead of a sheet or to attach some chafing gear; (c) that if by any chance you *should* lose your footing, a single 18-inch life line may serve only to trip you up and spoil what otherwise might have been a perfect dive.

Nordlys has always been equipped with triple life lines which completely enclose the deck. These life lines consist of $5/16$-inch stainless steel wire, parceled, served, and painted white to make them readily visible at night. The top line is 36 inches above the rail which in turn is 10 inches above the deck. Theoretically a man weighing 200 pounds could fall against these life lines from any part of the deck without danger of disaster. But in April, 1937, the stanchions were too far apart, the lines were slack and my unconscious form folded between the rail and the lower line like the proverbial sack of meal. We have since added extra stanchions spaced closer together, and spliced in a series of turnbuckles to keep the lines taut. It would be impossible now to go through the life lines, and, because of their height, somewhat of an acrobatic feat to go over them. For the helmsman, on dark and really breezy nights, we have provided a window-washer's belt into which is spliced 6 feet of $3/8$-inch line with a heavy snap hook on the end. When this hook has been snapped into an eye on deck, we are sure that our helmsman, at least, will stay with us.

The more adventurous crew members, like the life line skeptics, sometimes question the need for this device. A helmsman attached to his ship through the medium of a life belt seems to them just a tiny bit unmanly. The best possible answer is an understanding but firm persistence on the part of the skipper. Within 6 feet of the helmsman, hung on stanchions on either quarter, are two large-sized life rings painted white, each equipped with a water light. These life rings can be freed instantly by a quick pull on the single reef knots that hold them. On the side of the wheel box is a smaller ring to which, when we are at sea, 200 feet of light line is attached.

If in spite of such precautions, a shipmate manages to get through the life lines, or falls off the bowsprit, or stumbles over the counter on his way back from reading of the log, three things become immediately and vitally important.

First of all, one man must take charge and there must be no questioning his orders or his responsibility. There may be ten good ways to get the man back on board, but if an attempt is made to put all ten into effect at once, the result may be disastrous. (It's not a bad idea to appoint a *second* in command before you start your cruise. Then if the skipper himself goes over the side, he will not leave a leaderless group behind him.)

Secondly, a life preserver with a night light should be thrown at once. Be sure the night light is attached properly, so that the sealed opening in the end will be quickly broken and salt water allowed to mix with the chemical. Be sure, too, that the night light has at least 6 feet of line so that the man who swims to it can hang onto the life preserver without getting burned by the chemical.

The night light is important even in the daytime, for the smoke will help the swimmer locate the life ring even though a considerable sea is running. His eyes are close to the surface, and the life ring has only three or four inches of free board. In a moderately rough sea he may not see the life ring at all, if it is separated from him by even two or three waves. On the expedition that your author remembers so vividly that April day, a heavy sea was running. Three life preservers were thrown to him, all within 300 feet of where he was swimming. Yet he saw only one, and in that case the smoke from the night light was responsible.

Just one more caution on that life preserver. In your effort to throw it near the unfortunate shipmate, be sure you do not hit him with it! A direct hit with a heavy ring might make a good comedy situation for Hollywood, but it would very probably add to your immediate problems.

Thirdly, the skipper should instantly appoint at least one crew member to the sole job of keeping an eye on the man in the water. A yacht under sail may be moving at nine or ten knots and the man in the water will be soon well astern. A swimmer's head is a small object.

No matter how short-handed the yacht may be, one man *must* be spared from the problem of handling the ship and told to keep his eye relentlessly on the man in the water. At night the speed with which you get your life preservers with the attached night lights overboard may be the biggest single factor in getting your man safely back on board. A Very pistol with a white flare will light up a wide area of ocean, and for this reason, as well as for use as a distress signal, a pistol of this type should be readily available on every yacht that goes to sea.

What should the helmsman do when a man goes overboard? Seven out of ten yachtsmen will answer without much hesitation, "Jibe the ship immediately."

To this broadly accepted theory, I would like very humbly to take exception. It may be the most dangerous generality that has ever been bounced around a yacht club luncheon table. It originates, I believe, with yachtsmen who have been principally concerned with small boats where light sails are a minor problem and a jibe a casual maneuver.

The helmsman of an oceangoing yacht who jibes his ship with the wind well aft and a spinnaker, mizzen staysail or golly wobbler set, may easily turn a difficult situation into a tragedy. In the process of a quick jibe he may smash the main boom. (That happened, I believe, on *Hamrah*) His light sails will become a confused mess of canvas and rigging. As a result, the unfortunate gentleman who lost his footing may find himself well to windward, with his ship, in a relatively helpless condition, drifting away from him.

If the wind is well aft when the man goes overboard, the helmsman should immediately bring the wind on the beam, carefully note his exact course and stay on it. The ship's speed will be cut immediately and the luffing light sails can be quickly brought into control. When the gear is reasonably well cleared away, he can either come about or jibe, reach back on the reverse course, and with a fair amount of luck, pick up the missing crew member with ease. The yacht, sailing between five to eight points off the wind, will be readily manageable. She can come up or bear off quickly as required.

If the yacht is on a reach with the wind approximately abeam, the helmsman can bring her about or jibe as quickly as the light sails (if any) are brought under control. The ship can then sail back on the reverse of her former course under easy control. If the yacht is close-hauled, she should be brought about as quickly as possible and sailed back with the wind aft of the beam. This particular situation should be the simplest of the three, because light sails will ordinarily be no problem.

A great deal of care must be used in coming up to the man in the water. Above everything else, don't tackle him as you would your mooring buoy back in Larchmont. In any kind of a sea, distance and speed are hard to judge. The safest way may be to come up slightly to windward. Then back your head sails (or drop them entirely), strap in the mainsail, and drift back, taking him in over the quarter.

Some member of the crew will very likely want to dive over the side to the rescue. If the man in the water is injured or unable to swim, this drastic action may be justified. In that case the swimmer should take a life preserver along or at least an oar. The end of a long line would be still better. But unless the man in the water is really helpless, an individual rescuing party of this kind will only result in an undermanned yacht maneuvering to fish two men out of the water instead of one.

How about the dinghy? Under ordinary circumstances you had better leave it alone. It is probably lashed securely on deck, and even though you can quickly cut the fastenings, some time will certainly be lost, time that might be better spent sailing your boat. Most dinghies, moreover, are small and if there is any sea running, there is a good chance of swamping.

The biggest reason why more men don't fall overboard lies in the relatively few yachts that sail off-shore. If you are going to sea, don't underestimate the danger!

If you are the skipper, it's up to you to plan every detail of your equipment, the life lines, life preservers, night lights, and Very pistol. Visualize every conceivable situation in advance, and decide as exactly as possible what you will do.

If you are going along on somebody else's boat, put yourself through a private drill of your own as you stand at the wheel at night. Consider what you would do if your shipmate, who is working on the jib up forward, should happen to go over right then.

Make up your mind that *whatever* you do, you will respect the responsibility that lies on your skipper's shoulders, and not burden him with unrequested suggestions, when and if that dramatic moment comes.

P.S. By the way, if you *personally* ever expect to go over the side, don't wear green oilskins. And black rubber suits aren't much more visible. Old-fashioned yellow oilskins with all their defects are much the easiest to see.

⚓

26

Ordeal

NEVIL SHUTE

*The scene is England at war. The first bomb had fallen shortly be-
fore midnight. Instantly the lights had gone. Huddled with his wife
and children, Peter Corbett had heard the crash of masonry, the
sirens of police cars. "I think it must be an air raid," he said un-
certainly.*

*The phone went dead. There was no gas, electricity, radio, news-
paper, or milk for the children. Water was scarce, the sewers burst.
Then at night the planes came, flying at great heights, taking no
aim, just dumping their stuff. When cholera broke out, Peter knew
he must get his family to safety before "joining up," and smuggled
them out of the quarantined city to his little yacht at Hamble.*

PETER CORBETT kept his yacht, the *Sonia,* at Hamble. She was not the
sort of yacht worn with white duck trousers. She was nearly forty years
old, a gaff-rigged cutter with a straight stem and a long old-fashioned
bowsprit, based upon the style of the fishing smacks belonging to the
east coast village where she had been built. Her hull was low in the water
and painted a dull black, her sails were tanned, her decks painted with
buff paint which made them tolerably watertight when the paint was
new and unbroken in the spring.

She was an aged, dirty little boat, not very sound, but Joan and Peter
thought the world of her. She was their hobby and their holiday, deep
laden with sweet memories of escape from their routine. As a permanent residence for two adults, two children, and an infant, her accom-
modation was not impressive. From the bows, she had a forecastle where
a water closet stood starkly between the chain locker and the cooking

galley. The galley was served by two Primus stoves, one of which carried a rusty tin cooking oven. A water tank of about fifteen gallons' capacity was clamped to a bulkhead; a little crockery was stored with the frying pans and saucepans in a cupboard. Aft of the forecastle the saloon was furnished with a settee berth on either side and a swinging table in the middle, with one or two small lockers beside the settees. A paraffin lamp in gimbals swung from the bulkhead. Aft again, one passed up on deck into the cockpit by means of a couple of steps forming the fore-end of the engine cover, removable to permit the flywheel to be cranked, knuckles to be damaged, or, occasionally, arms to be broken.

Corbett had paid two hundred pounds for her six years before. It was a lasting wonder to him that two hundred pounds could have bought so much happiness. As he drove his overloaded car through the streets of the malodorous, stricken city that morning, Corbett was not depressed. It takes a very little thing to lift the spirits of a man. He was leaving his home for an indefinite time, leaving his house, his business and his office, ruined and abandoned, flying with his family from death by high explosive or disease, journeying towards a future all unknown. And yet his heart was light. Routine was broken; there would be no more drafting of conveyances for a time, anyway. The sun was shining after the rain of the night. He had a hundred pounds in his pocket. And, above all, he was going to his boat.

They went to bed early, in preparation for a heavy day. They slept for an hour or so; then, punctually as an alarm clock, they were awakened by the raid. For a time they lay in their bunks listening to the concussions in the distance.

A salvo fell very close to them, on the marshlands behind Hamble. The children woke and cried a little; they got up to comfort them and turn them over to sleep again. They stood together in the hatchway for a while with blankets draped around them. Bombs seemed to be falling all over the countryside.

"I don't believe they're hitting Southampton at all tonight," said Joan. "They're rotten shots."

Corbett nodded. "They're getting wilder and wilder. I don't believe they'll do much damage with this raid, except by a sheer fluke."

Joan said, "They couldn't miss London like this, though."

"No, London's different. I bet they're getting hell up there."

After a time, before the raid was ended, they grew bored and cold, and went back to bed. "You'd better wake me if the ship begins to sink," said Corbett. "I'm going to go to sleep." Inured to the concussions by familiarity they fell asleep, stirring and turning over now and then at the nearer explosions. Presently all was still again, and they slept quietly.

They did not sleep for long.

Soon after midnight the raid began. They woke to the sharp crack of guns; there was an anti-aircraft battery located on the edge of the New Forest, not very far from them. The guns went on incessantly, monotonously; they had a sharp, piercing crack that hurt their ears. The children woke up, and began to cry.

"Hell," said Corbett. "Where's that cotton wool?"

They pressed wool into the children's ears, and into their own. They couldn't get any wool into the baby's ears, so they put pads of wool on top and bound the little face round with a bandage while the child yelled and struggled. Then they had done all that they could do; for a time they lay in their bunks listening to the detonation of the bombs.

Presently, exhausted by the whimpering of the children and the screaming of the baby, they got up and made tea, and sat in the saloon in the darkness with the children, drinking it.

Corbett said, "It won't go on much longer."

As he spoke, there was a rushing, whistling sound, and a great splash near at hand as something heavy fell into the water. What happened then was past description. The vessel seemed to rise bodily into the air beneath them, plucking at her anchor chain with a great crack that shook her to the stern. She was lifted, and thrown bodily onto the surface of the sea on her beam ends, with a crash. In the saloon they were all flung together in a heap on the low side, stunned and deafened with the detonation. On her beam ends she was carried swiftly sideways towards the centre of the channel; then she seemed to strike the bottom with her topsides, though she had been anchored in two fathoms. Slowly she rose till she was nearly on an even keel. Then a great avalanche fell upon her, smothering her down, pressing her underneath the tumult of the sea. A ton of mud and water poured down into the saloon through the half-open hatch; she was spun bodily around. Then she rose, streaming like a half-tide rock, and drifted out towards the middle of the channel.

Deafened and dazed, Corbett groped his way to the hatch and clambered out on deck. By some freak of chance the dinghy was still with them; sunk to the gunwales, she was still attached to the stern by her painter. The boom was trailing in the water, topping lift and mainsheet carried away. There was a tangle of loose gear at the foot of the mast that he could not stop to investigate; the glass of the cabin skylight was shattered. The anchor chain hung straight down from the bow, broken off short; the vessel was slowly rotating out into the middle of the channel. She was much lower in the water than usual; the decks were deep in slime.

He hurried aft to the sail locker, got a line, and bent it to the kedge anchor. Then he went forward and anchored her roughly with the kedge and warp; she brought up in about six fathoms. Coming back aft, he saw that Joan was in the cockpit, working at the pump.

"Are the children all right?"

"I think they are. Look, take over pumping, Peter, and I'll go and see to them. There's over a foot of water in the cabin."

He went to the pump. "Mark the level in the cabin, and tell me if I'm getting it down at all."

He settled to the pump. In the cabin he could hear her sloshing about in the water, could hear her comforting the children. Presently he heard the roaring of the Primus stove. He pumped on steadily. On shore the battery was still throwing its barrage to the sky; bombs were still falling round about. At the end of twenty minutes Joan said, "You're getting it down, Peter. It's an inch lower than it was—an inch to an inch and a half."

He rested for a minute, and began again. Presently, having soothed the children, she came to him with a cup of Bovril; he drank it gratefully while she relieved him at the pump.

He asked, "Do you think she's making water?"

"I don't believe she is. The level's going down all the time. I think it's only what came into her by the skylight and the hatch."

"Lord," he said, "we don't want another one like that."

"What about that Quarantine anchorage, now?"

"They can keep it."

He busied himself with the boom. When that was inboard he went round the deck assessing the damage. It was not so bad as he had feared. The little yacht was injured, but she was not incapacitated; there was nothing there that he could not patch up and repair himself, given the time. He went aft and pulled the sunken dinghy up to the counter. Joan left the pump and went to help him; together they hauled it out of the water, emptied it, and put it back afloat. Then Corbett went back to the pump, and Joan went down below.

"The water's practically off the floor," she said. "I don't believe she's leaking more than usual. I'm going to change the kids into dry things."

An hour later, the pump sucked. Corbett went below, exhausted and with a violent headache. He poured himself out a stiff whiskey, and gave one to Joan. "We'll get away from this bloody place as soon as we can," he said wearily.

"Is the boat all right to get away?"

"I think so. I'll have to go and find the anchor. But it's got a buoy on it."

He made her lie down on the other settee. Then he changed into dry clothes and put on his oilskins, spread a sail doubled over Joan's sopping bunk, pulled the wet blankets over him, and fell into a heavy sleep.

When he awoke, three hours later, it was daylight. He got up stiffly and took off his oilskins; Joan and the children were still sleeping. He went on deck, got a bucket, and started to swill away the slime that covered the vessel.

The morning came up sunny and bright. Joan heard him moving about on deck, got up, and came to the hatchway. She wrinkled up her nose at the mess on the deck; then she went back and started to get the children up. Corbett went off in the dinghy, found the anchor buoy, and raised the anchor with ten feet of broken chain attached to it. He took it back on board and shackled it onto the remainder of the chain.

A couple of hours later they had more or less recovered from the incident of the night. They had had a good meal and had washed up; their clothes, their blankets and their bedding were laid out on deck and drying in the sun. Corbett was drying the magneto of the engine in the oven, and Joan, with sail needle and palm, was repairing a long slit in the mainsail.

They worked all morning in the sun; by noon they were ready to get under way.

27

Crossing a Grim Atlantic

ERLING TAMBS

AMONG the numerous and varied types of small craft constructed by maritime peoples in all parts of the world to resist the fury of the waves off bold and rocky shores, I believe there is no type more seaworthy than the Norwegian double-ender. A sailing vessel deserving the name was never created as the result of a single man's genius; she is the outcome of seafaring experiences of the generations which stand behind her, and her seaworthiness may be said to be in proportion to the number of such generations and the severity of the circumstances with which she has to contend.

Considering that the conditions of navigation on the rockbound coast of Norway are exceptionally hard, it is only natural that the craft used by our pilots and fishermen are particularly fit to cope with seas and gales. However, on rare occasions it may happen that the ocean rises to such violence that even the best of man-made ships will find themselves in serious danger.

When I read about the endeavors of the Cruising Club of America to arrange an ocean race from Newport, R.I., to Bergen, Norway, I was enchanted. It was in November and I took it for granted that there would be just time enough for Norwegian yachtsmen to build a boat or two to take part in the race, although I realized that our boats would have to leave Norway about the middle of March in order to arrive at Newport in time for the start, the 8th of June.

Unfortunately it proved that our boat yards had orders on their hands which made it impossible for them to guarantee delivery as required and thus it looked as if our country would not be represented in one of the greatest yachting events ever staged. It seemed as if Norway would be reduced to the rôle of taking the time when an international crowd of yachts came to finish a race in our country. I admit that I did not like the idea.

Nevertheless, in January I chanced to come across a boat to suit my purse—and to some extent—even my requirements. To be sure, she was not a yacht and neither was she built for racing, but at least she was sturdy and seaworthy, two qualities which, under certain circumstances, might outweigh even the smartest rigging and the finest lines. Yet I dare say that no one will suspect me of having harbored any hope of winning the race. My participation in the race with *Sandefjord* was to be mainly representative—to show the Norwegian flag and our honest intentions—and, perhaps, to be considered in the light of my hope that it might form the humble beginning of our endeavors to assert ourselves at a game to which we seem to have a birthright.

Sandefjord is a former life-saving boat, 46 feet 8 inches in length over all, 16 feet beam and 7 feet 6 inches draft. She is 22 years old, in good condition and very heavily rigged.

We sailed from Sandefjord, the town after which my boat had been named, on the 14th of March with the Canary Islands as our destination. I had chosen to call at Las Palmas for the purpose of putting my boat in yacht-like trim. In winter the weather conditions in Norway are not in favor of such work as painting and varnishing.

We were prepared for a rough time in the North Sea and the Bay of Biscay. However, in the North Sea calms prevailed and the Bay of Biscay we crossed in less than 60 hours with a fine fair wind from the north. Yet our passage became a lengthy affair on account of subsequent long spells of calm off the Portuguese coast. We arrived at Las Palmas on the 15th of April and left again a week later, with Newport as our destination. It was on the morning of the 22nd of April.

Our start was promising. Six hours after leaving Las Palmas we passed Maspalomas, the southern point of the island of Gran Canaria, and a week later we had covered the first thousand miles of our journey. But from then onward the Northeast Trades proved capricious, blowing sometimes from the south and sometimes from the north and then again leaving us becalmed for three or four days at a time.

Working steadily westward about the latitude of 21° N, we reached longitude 55° W on May 12th and gradually altered our course northward. The wind was light from the south but presently attained sufficient strength to give us fair headway with all sail set. The barometer was high and very steady until the night of May 16th, when it showed a slight fall, which became more marked as the night went on. At the same time the wind rose to Force 6 which induced me to stow the balloon jib, the gaff topsail and a water sail of my own invention, which we carried under and to leeward of our mainsail. Under the remaining canvas (forestaysail, mainsail and mizzen), we carried on, steering NW½N with the wind on our port quarter. It was brisk sailing.

At noon of May 17th the wind had increased to Force 7 or 8 with frequent fierce squalls, compelling us to keep the boat's head two or three points off her course. The sky gradually took on an ugly, threatening appearance and it soon became evident that we were in for foul weather. At 3:00 p.m. we furled the mizzen and double-reefed the main and staysail. Attempting to set our storm jib, this sail carried away and knocked a hole in the staysail, thus rendering both our heavy weather headsails unfit to heave to under.

Leaving the staysail set, I took the soaking wet storm jib down into the cabin to make the necessary repairs. It was a slow, strenuous job and, while attending to it, I kept an eye on the barometer, which had begun to drop with great rapidity. Now and then I would take a look on deck. The surrounding seas were gradually turning into mountains as they were lashed by the shrieking wind. Yet I felt no apprehensions regarding the safety of the boat. At 7:00 p.m. it became necessary to have two men at the tiller. They were Kaare Tveter and Thorleif Taraldsen. There was no time for supper on that night. However, the storm jib would presently be repaired and then it was my intention to heave to.

At half past seven the jib was mended and ready for service. We were just preparing to take it on deck when Kaare Tveter, at the top of his voice, shouted down the hatchway: "Skipper! Come on deck; we cannot manage her."

We left the jib below and rushed on deck. The violence of the wind was terrific. It was no longer a storm, it was a fullblown hurricane. The boat rushed thundering through the tumultuous foam, while spray and Sargasso weed hit our faces with maddening blows and the driving rain cut like needles. A hasty glance at the compass showed me that the wind had hauled around to the southwest. We were heading NE by N. I shouted to the boys at the tiller: "Watch out that she does not jibe, but whatever you do, don't broach to!"

During the last half hour the sea had changed entirely. To be sure, it had been mountainous before, but more in the manner of good-natured giants, who might do harm if one did not look out. Now the sea was riotous, as if those giants had turned raving mad, tumbling about in a vicious, drunken manner, suddenly joining to pile up huge towers and form tremendous combers where one least expected them. I have never seen a sea so threatening.

We were, however, too busy to look about us much. The three of us, Peter Archer, Einar Tveten and myself, hauled down the staysail and tried to make it fast, but the gaskets broke. While I was hanging on to the sail forward with Einar Tveten beside me, I sent Peter Archer amidships to fetch a length of heavy rope from under the life boat, where we kept stowed a variety of gear. Just then the boat stuck her nose under and I jumped up and grabbed hold of the forestay, while the water surged around me up to my belt. Even then I did not realize that there was any immediate danger. A quick glance showed me that Einar was hanging onto the rail close by, while Peter—on hands and knees—was on his way forward with the gasket. None of us three saw the wave that suddenly rose astern. But Thorleif Taraldsen, who was at the tiller, saw it and he swears that it was as high as the masthead.

As for us, we did not know what was happening, when all at once the boat made a second plunge. In a fraction of a second the water rose about me, above my head, still further, and I felt myself being dragged under in the seething foam by what I believed to be the suction caused by the sinking boat. I could not think of any other explanation but that the planking had sprung from the stem and that the *Sandefjord* was going down with all hands. I had lost my hold on the stay; how, I do not know. I kept on sinking, sinking through water white with foam, and I did not fight. What use could it be to fight for a short few minutes of life, when death seemed to be certain after all? What chance could a swimming man have in such a sea? I swallowed sea water.

Perhaps it is not so easy to reason one's self out of life. Be it that nature will fight to the end or that I suddenly realized that my children were about to lose their father, I swam to the surface with all my power. But I managed to draw only one breath of air before I was pulled under again, and then I gave it up. Why prolong the pain with futile efforts? So far, I had seen nothing but foam about me. However, I was shot to the surface again, and this time I saw the boat, perhaps 25 yards to windward.

She was sailing slowly by on the starboard tack, driven by a minute rag left from her mainsail. I saw no one on board but heard voices and called out for a life belt. It was thrown, but I could not find it and instead swam for the boat, which I managed to reach before she had sailed by. Somehow, it seemed as if the sea which had swamped us had smoothed the surface of the water around.

Thorleif Taraldsen was the only man left on the ship. He had been jammed in the cockpit by the mizzen rigging, the mast having carried away, breaking about two feet above the deck. He helped me on board. At the same time Einar Tveten and Peter Archer came back on board, the former being washed in over the stern and catching hold of the mizzen traveler, the latter climbing over the mizzen spars, which, still hanging by one of the shrouds, floated under the starboard quarter. Peter Archer's experience in the water had been almost identical with mine, except that he came to the surface on the opposite side of the boat, about 25 yards to windward. Kaare Tveter never returned.

Back on board I saw the condition the boat was in. I felt almost sorry that I had not drowned in the first instance. The *Sandefjord* certainly did not look as if she could float long. In the cabins rusty water splashed about, two feet deep. Bedding, clothing, crockery, provisions and various personal belongings littered all the floors below; the deck was swept clean, the life boat, binnacle and compass and much gear having been washed away; the rails were smashed, the mizzen rigging lost and the main rigging apparently about to go overboard at any minute. The port anchor had broken loose and disappeared with part of the gunwale. The starboard lightboard had been lost, the starboard shrouds were partly broken, the chain plates were all twisted and the forestay—the only stay holding the rigging forward—had given way to such an extent that I felt doubtful if it could be used to stay the rigging again. To all appearance the boat was a total wreck, about to founder.

But the worst of it all, and the main cause of our almost apathetic depression, was the knowledge that Kaare Tveter was missing and would never return.

The accident happened about 26° N and 63° W, in a part of the ocean where ships rarely go. My first object, after clearing up the ship as well as our scanty resources would permit, was to reach the nearest steamship lane in case we should require assistance, because at that time it did not seem probable that our remaining rigging would last many days. I decided to sail south, regain the Trades and then sail west. Fortunately, after the storm, which abated the morning after our disaster, we had a spell of fine weather, which gave us a chance to repair the rigging and invent makeshifts for much of the gear which we had lost.

By slow degrees we managed to get the boat in fair trim, although the best we could ever do in a head wind was to sail at right angles to the wind. We had lost our mainsail, staysail and mizzen and had no spares, but we managed to turn the remnants of the mainsail into a useful try-sail and, by rigging a spar to the main boom end and using a jib as a watersail under the boom, we could do fairly well with the wind abaft the beam.

Gradually we regained our confidence and, when finally we met the American tanker *Beacon* in 27° N and 70° W, we asked nothing but to be reported; we had made up our minds to sail to Newport, R. I., our original destination, without help.

We arrived there on the 12th of June, 52 days after our departure from Las Palmas and 26 days after our accident. We came four days too late to see the start of the ocean race to Norway in which it had been my intention to take part, but the friendly welcome and the assistance given us by the residents of that most hospitable city helped us to forget our disappointment and amply rewarded us for the trouble we had taken in getting there.

⚓

28

Sailing Alone Around the World

A. ROMNEY GREEN

To BEGIN at the beginning, Captain Slocum was a retired merchant skipper who had owned the last ship he commanded, and lost her on the Brazil coast. His voyage home to New York with his family was made, he says, "in the canoe *Liberdade* without accident." Being apparently short of a job, he proceeded to build the famous *Spray*. He says distinctly that she was "built over" entirely by his own hands from a very old hulk of which he had been made a present. "She changed her being so gradually that it was hard to say at what point the old died or the new took birth." Yet the *Spray* had a new keel and new stern post, oak trees felled by the gallant captain's own "ax;" new timbers, straight saplings steamed and bent; new planking and new deck of 1½ in. Georgia pine. Where, then, did the old *Spray* come in? And how comes it that so tight and sound a ship as the *Spray* proved to be should have been largely built, as she apparently was, of absolutely green wood?

The *Spray,* presumably the new part of her, cost, we are told, $550 for materials, and was built by the captain's own hands in thirteen months, he then being, from what I remember of his appearance, about fifty-five years old; a bearded, iron grey, wiry-looking man of medium height. She was 37 ft. long by 14 ft. beam, and the building and sailing of such a ship single-handed by a more than middle-aged man of her skipper's modest dimensions is in itself no mean achievement.

Other statements which have been found difficult of belief are those regarding the wonderful self-steering qualities of the *Spray,* in virtue of which Slocum was usually able to leave the helm for hours or even days together. He sailed, for instance, from Thursday Island to Keeling Cocos, 2,700 miles as the crow flies, in 23 days, during which he spent only three

Reprinted by permission of *The Yachting Monthly*.

hours at the helm, including the time occupied in beating into Keeling Harbour. Again, sailing with a friend in the mouth of the Plate, "Howard sat near the binnacle and watched the compass, while the sloop held her course so steadily one would have declared the card was nailed fast"—this "in shoal water and a strong current."

Yet again, Slocum avers that he carried no chronometer. He had a good one, but "it would have cost fifteen dollars to repair and rate it"—quite out of the question! He calculated his longitude, apparently with great success, from a rotator log and occasional lunar observations.

Let us suppose, however, that Slocum really had a chronometer covered up somewhere in his cabin; that he did exaggerate the *Spray's* self-steering qualities, and that the story of her building is a mere fabrication. The indisputable fact remains that a certain Captain Slocum sailed alone 'round and more than 'round the world—he crossed the Atlantic three times—in a certain vessel of which the lines were afterwards ascertained by an American yacht club. The archives of numerous provincial newspapers all 'round the world contain testimony to this effect; corroborated in each case by the personal curiosity of hundreds of sightseers, of which, at Durban in November 1897, I was one.

I had already read the account of Slocum's arrival at Durban in the local paper, which was handed to the worthy skipper on landing, with an American promptness which must have made him feel quite at home; informing him that he had been sighted at a distance of fifteen miles from the Bluff Station, and that at a distance of eight miles he had reefed his mainsail in ten minutes single-handed. And I only wish I could have seen him tackle the dangerous harbour bar and entrance, preceded, but not towed, by the pilot tug, in the heavy sea that was then running.

On the afternoon, however, of the following day I got on board a little home-made craft of my own, an 18-ft. canoe yawl of diamond section, and looking for what I then conceived as the "limit" in the way of a single-handed sailing boat, I was surprised at the size, and especially at the immense beam and solidity of the *Spray* as I found her at her moorings, the girth of her mast, the weight of all her spars and tackle. The skipper was on board, and though his demeanour was not encouraging, I luffed my diminutive craft up alongside, and was brusquely allowed to make fast and come aboard the monster. Being young at the time, and not knowing that Slocum was a poor man, my natural delicacy did not permit me to realize that this should perhaps have been a business proposition—I believe sixpence was his standard charge for inspection of the *Spray* to those insignificant persons to whom he did not bring an introduction.

One of the outstanding features of Slocum's voyage is, indeed, that it was that of a poor man without government assistance. Columbus must wait for a royal subsidy to discover America, and subsequent explorers have usually needed similar endowments. It is true that they went first, but it is almost equally remarkable that Slocum went alone, and it was on his own bottom that he undertook his great adventure.

As to the literary quality of Slocum's work, here is the description of the close of the first day of his tremendous voyage:

"I made for the cove, a lovely branch of Gloucester's fine harbour, again to look the *Spray* over and again to weigh the voyage, and my feelings, and all that. The bay was feather-white as my little vessel tore in, smothered in foam. It was my first experience of coming into port alone, with a craft of any size, and in among shipping. Old fishermen ran down to the wharf for which the *Spray* was heading, apparently intent upon braining herself there. I hardly know how a calamity was averted, but with my heart in my mouth almost I let go the wheel, stepped quickly forward, and downed the jib. The sloop naturally rounded in the wind, and just ranging ahead laid her cheek against a mooring pile at the windward corner of the wharf so quietly after all that she would not have broken an egg. Very leisurely I passed a rope around the post, and she was moored. Then a cheer went up from the little crowd on the wharf. 'You couldn't 'a done it better,' cried an old skipper, 'if you weighed a ton!' Now, my weight was rather less than the fifteenth part of a ton, but I said nothing, only putting on a look of careless indifference to say for me, 'Oh, that's nothing' for some of the ablest sailors in the world were looking at me, and my wish was not to appear green, for I had a mind to stay in Gloucester several days. Had I uttered a word it surely would have betrayed me, for I was still quite nervous and short of breath."

But on the whole Slocum is too chary of matter even so little technical as this. Himself a professional, he does not seem to realize the delight of the amateur in every detail of the grand profession. He only mentions incidentally his use of a sea-anchor in the last and worst of the many great gales through which he passed. In another bad gale he scuds under a fore-staysail only, sheeted flat amidships, and towing a hawser over each quarter to break the great Pacific combers. But he has no hard and fast theories and recipes like those of Captain Voss with his sea-anchor, Knight with his whaleboat stern, or T. F. Day with his diamond section. He has scarcely even a remark to offer on the relative merits of the sloop and yawl rigs, though he starts with the former, and takes considerable trouble, less than halfway through his voyage, to change to the latter, as its champions will be glad to hear. The only dogma of which he delivers himself is damnation to overhangs: "For your life build no fantail overhang on a craft going off-shore."

But though Slocum is not prolific of theories, his practice is often instructive, and it is especially interesting to find in the *Spray,* as in the coble, the Thames barge, the Dutchman and the skipjack or sharpie, a witness to the seaworthiness and efficiency of the skimming dish type —for it will be seen that she was little more than a heavy skimming dish.

With regard to the alleged danger of capsizing this type of boat, it is interesting to find that Slocum's only fear was for his mast and rigging. On one occasion a terrific squall in the Azores parted one and stranded another of his lanyards. Curiously enough he never speaks of heaving to, as presumably he might have done on the other tack in this instance; but he "rounded close under a bluff" to repair the damage, and was immediately boarded by a Customs officer, who took him for a smuggler. The broken lanyards, however, "turned the incident in his favour," and the crew helped him to set up his rigging. "I have always," he says characteristically, "found this the way of the world. Let one be without a friend, and see what will happen."

He suffered from loneliness at the outset, but this soon wore off, and on one occasion, after sailing alone for forty-three days without a port, he might have touched at Nukahiva, but he actually preferred to hold on another twenty-nine days for Samoa. And here is a description of his life on board in the Trade winds:

"The *Spray,* under reefs, sometimes one, sometimes two, flew before a gale for a great many days with a bone in her mouth, toward the Marquesas in the west, which she made on the forty-third day out, and still kept on sailing. My time was all taken up in those days—not by standing at the helm; no man, I think, could stand or sit and steer a vessel round the world: I did better than that, for I sat and read my books, mended my clothes, or cooked my meals and ate them in peace. I had already found that it was not good to be alone, and so I made companionship with what was around me, sometimes with the universe and sometimes with my own insignificant self; but my books were always my friends, let fail all else. Nothing could be easier or more restful than my voyage in the Trade-winds.

"I sailed with a free wind day after day, marking the position of my ship on the chart with considerable precision; but this was done by intuition, I think, more than by slavish calculations. For one whole month my vessel held her course true; I had not the while so much as a light in the binnacle. The Southern Cross I saw every night abeam. The sun every morning came up astern; every evening it went down ahead. I wished for no other compass to guide me, for these were true. If I doubted my reckoning after a long time at sea I verified it by reading the clock aloft made by the Great Architect, and it was right.

"There was no denying that the comical side of the strange life appeared. I awoke sometimes to find the sun already shining into my cabin. I heard water rushing by, with only a thin plank between me and the depths, and I said, 'How is this?' But it was all right; it was my ship on her course, sailing as no other ship had ever sailed before in the world. The rushing water along her side told me that she was sailing at full speed. I knew that no human hand was at the helm; I knew that all was well with 'the hands' forward, and that there was no mutiny on board."

This between his recurrent spells of danger and hardship and almost incredible labour, was surely a life worth living; at all events for the kind of man who combines the philosopher and the athlete, the hermit and the adventurer, in such just proportions. But how many athletes and adventurers would enjoy, how many philosophers and hermits would pay the necessary price for, such a solitude?

The sailing man is often asked—I was going to say by the Philistines, but let us say merely by the uninitiated amongst his friends and acquaintances—what pleasure or profit he can possibly find in the extreme discomforts and occasional dangers of his holidays on a small cruiser. And he generally makes the mistake of trying to meet them on their own plane. He says perhaps that the fresh air and exercise is good for his health, and that the danger and hardship is good for his character, and that health and character are both good for his business.

29

Nautical Measurements

Like the metric system, nautical measurements have an origin which is by no means empyrical. Like the metric system, also, they are based upon the dimensions of the earth.

The *knot* is a unit of velocity, and not one of length. It is one *mean* sea mile per hour (a sea mile is not constant, but varies from 6045.94 ft. at the equator to 6107.85 ft. at the poles). At any place it is a minute of latitude there. Obviously, it would be inconvenient for a knot to vary, so the arbitrary figure of 6080 ft. per hour is accepted as the knot in Great Britain. In France and Germany it is 6076.23 ft. (1852 metres) per hour, a closer mean to the nautical mile. In the United States 6080.27 ft. per hour is the standard. The origin of the term knot explains the term itself—originally a ship's speed was gauged by a long line attached to a floating log, and knots were made in this line at fixed intervals, and their number was noted as the line ran out over the taffrail.

To ascertain a vessel's speed she is driven over what is called a "measured mile," of which there are many such round our coasts. On shore two pairs of tall spars are erected 6080 ft. apart. One starts the run by timing when the first pair are in line, and again timing when the second pair come into line at the finish, the observation being taken from a fixed point on board. The calculation is a matter of simple proportion: if the vessel travels one sea mile in so many minutes, she will do so many in 60 minutes, and that will be her speed in knots.

All nautical measurements are based on the "Parallel of Latitude" and the "Meridian of Longitude."

A *Parallel of Latitude* is an imaginary circle drawn round the earth, taking a pole as the centre. Obviously, the equator is the largest possible, being the greatest circumference.

Reprinted by permission of *The Yachting Monthly*.

A *Meridian of Longitude* is an imaginary line drawn from pole to pole, and cutting the equator and all other parallels of latitude at right angles. Owing to the comparative flatness at the poles this is not a perfect semi-circle, but it is treated as such and divided into 180 parts (half of the 360 parts or degrees into which all circles are divided). Owing to the polar flatness these parts or degrees are not equal, being shorter near the equator and longer near the poles. Thus the 180 parts or degrees of latitude into which a meridian or line of longitude is divided increase in length from the equator to the poles. Similarly, the sub-divisions of the degree into 60 parts or minutes, and these minutes into 60 parts or seconds, increase in like proportion as the degrees.

Parallels of latitude, being circles, are divided into 360 equal parts, and these are called *Degrees of Longitude,* and a semi-circle giving the same degree of longitude on every parallel is a *Meridian of Longitude.* Each degree of longitude is divided into 60 minutes, and each minute into 60 seconds.

If, therefore, we want to find the nautical or sea mile at any place it is the length of one minute of latitude there. At the equator it is 6045.94 feet, and at the poles 6107.85 feet. Except in the arbitrary acceptance of the term (6080 feet) in defining the distance factor in the knot, the sea mile varies from the equator polewards.

A nautical or sea mile must not be confused with a *Geographical mile,* which is a minute of longitude on the equator—6087.1 feet. It is certainly *nearly* the same length as a nautical mile at the equator, but not quite. It is, however, a fixed measurement.

Degrees, minutes and seconds of longitude diminish in length from the equator proportionately as the circles of parallels of latitude themselves diminish. All these points may be followed more clearly with the aid of a chart where on the lines of longitude and latitude are drawn.

For shorter nautical measurements the mean nautical mile of 6080 feet is adopted and divided into 10 equal parts called *Cables.* For convenience these are not 608 feet but 600 feet or 200 yards. Again a cable is divided into 100 *Fathoms* of six feet or two yards each.

The statute mile, 5280 feet, must *never be* confounded with either the nautical or geographical mile.

NOTE:

One degree is written thus ...1°
One minute   "   "   ....1′

One second is written thus....1″

The following then are the nautical measurements:—

1 geographical mile = 1′ of longitude at the equator (6087 feet).

1 nautical or sea mile = 1′ of latitude (mean length taken as 6080 feet).

1 nautical or sea mile = 10 cables.

1 cable = 100 fathoms.

1 fathom = 6 feet.

(A cable is accepted as 100 fathoms in all latitudes.)

Thirty-One Hundred and Fifty Miles

BRUCE AND SHERIDAN FAHNESTOCK

WE WERE heading out into what once was called the Mysterious Sea. Mariners were once afraid of it, believed great serpents and beautiful sirens occupied it, thought nothing but the most violent storms racked its deceiving blue surface. We were heading out into the part of the world where the beaches are white, not black lava, where the sea is fringed with green, not scraggly brown bush.

Just as we passed the northern point of Narborough we saw a wreck— an old fishing boat driven up among the jagged lava rocks. We searched the land with the glasses but saw nothing—no sign of life. We itched to go ashore and explore, but our passage to the Marquesas had been long enough in getting under way. We still had delusions about a record passage. We optimists aft, with the exception of that old grouch Puleston, told ourselves the run across the 3,150 miles of water ahead of us couldn't take over twenty-five days.

But what did the wild wind say? Nothing! The mythical Trade which for some reason stuck to *Svaap* on her fast passage bleweth not for a ship that needed wind and plenty of it. The first night out a heavy fog settled over the sea. It was the first and last fog of the entire voyage in *Director*.

Our point of departure from the Galápagos for the Marquesas on the 27th of June, 1935, was ten miles south of the Equator. On the following noon we found we had made eight full miles in the desired direction but had drifted twenty-five miles north. And this while we held a course of west by south. That was the beginning of our battle to regain the Line. The strong nor'east set made it a long job. When asked, "Have you ever crossed the Line?" we answer modestly, "Not more than twenty times." It is true: by day the wind was sufficient to help us south across, by night the current was so strong that it set us back up again, and so it went. We covered only 860 miles in the first two weeks of sailing.

We had time to think of many things. But, as is the custom aboard ships, we thought most of food and things to eat! "Come and get it!" and an undertoned "before I throw it out!" never ceased to be the most important and enjoyed command aboard *Director*. At seven the watch called all hands with "Rise and shine, you grimey blokes!" or some such endearing term as a delicate awakener. For an hour or so preceding this, the other member of the watch had been fiddling about in the galley with the hatches and vents so arranged as to waft the delicious flavors aft to the sleepers' nostrils. It was fairly evident someone had to offer some redeeming feature for a rough rousing. Hot coffee, hardtack, and jam went a long way in muffling the grumblers in their just-awakened irritability. On the second cup of coffee and the first half-pound of jam the world began to clear.

So our day would begin correctly, with the backbreaking task ahead of one—the backbreaking task of reading idly until twelve. Sheridan would be at us every moment with rough commands, such as: "Read this book!" or "Take a nap!" or even, "Move over to starboard; there's less sunlight over there!" Yes, these Corinthian skippers are a hard lot.

The busy morning would disappear when punctuated by another summons from the galley. Lunch. Lunch over, we would again repair to the deck and wear ourselves to the bone turning pages, gripping fountain pens, or, if the worst should befall us, we might squat upon the wheelbox for a time while the ship steered herself.

Everyone cooked. And the results were astounding. Sometimes overpowering! Breakfast was done by the pair who had the morning watch. They also cooked the supper of the evening before with no more success than they had with their breakfast. The graveyard watch always cooked lunch, and the twilight, dinner. As the watches advanced—they went forward a watch every night—so did the cooks. (Mind you, we speak not of quality.)

The competitive plan of cooking has great faults. Abroad *Director* in the first few weeks out of port, the rarer of the ship's delicacies began to go, almost as if by magic. Out of the galley would come a terrible oath:

"What —— dudgeon stole that —— boned chicken!" and then "I'll split him over the spring stay!" But the boned chicken, highly camouflaged, was in a none-too-successful hash that stuck to the pan in the watch before and was tossed, on the quiet, overside. Thus went on a never-ending sneaking about with noses in cookbooks—an occasional "Hey! Stop reading over my shoulder! Find your own recipes!" Thus began the unending search for new aliases of beans and hash.

During the long afternoons one Jack could be heard saying to another: "The Waldorf book says three onions and a bit o'chopped parsley!" And his galley partner would answer: "But Oscar isn't here and the Fanny Farmer book says three tablespoonsful of chopped onion and a pickled bay leaf." And each afternoon the galley crew changed, but the argument was almost the same.

Once, and we hope it never happens in his life again, Denny mislaid his special cookbook. For days he was quiet, for days his face had a drawn, pained look. We searched everywhere. No book. Dennis was close to breakdown. How could he cook without that marvelous book of formulas! Suddenly the book was found. Dennis dashed below, rattled things about, worked for hours. When he rang the Brion Bell we fell below. There on the table was the same gray stew he had made a dozen times! The same grayish mess he had made since his cooking had begun. But Denny was happy and we all ate heartily of what he described as a "typical English dinner."

In the afternoon, when all thought was off the ship, far away in a favorite restaurant somewhere on a distant shore, wrapping itself in imagination around fabulous steaks and unheard-of chops, the galley crew would be driven below at a rope's end to prepare the evening meal. A tremendous clatter would be set up. Then one of them would shout: "Some one of you so-and-sos open the galley hatch!" That meant rising temperature; that meant food couldn't be long in coming. It also meant the primus stove wasn't primus-ing properly.

And in that word "primus" you have the reason for most of the really well-thought-out cursing that took place aboard *Director*. After chucking off our hot coal stove in Panama, we had come out into the Pacific with two pressure kerosene primus stoves set in gimbals. The gimbals were fine. The stoves themselves looked very shipshape when their brass was burnished. But they were rotten stoves. We repaired them, nursed them, bought new parts for them, pumped them up fifty times a meal, cursed them, hated them for two and a half years. There was always the sickening smell of oil about the galley no matter how much scrubbing poor Hey Hey did. There was always a part falling off of one of them just as a duck was reaching his baking crisis, and there was never a moment when we could be sure those diabolic monsters wouldn't flare up and set fire to the ship! We found, by chance, that the chandler who had charged us twenty dollars for the stoves had substituted an inferior make of stove for the Swedish ones we had thought we were getting.

At last, those starving on deck would hear a welcome ringing and shouting from the dungeon below. A huge kettle of potatoes would steam to greet us. Fried bananas—when we had them—spinach, string beans floating in butter, taro, yam, and other things too numerous to mention would be heaped upon the edged table so that there was no room for plates. In silence the meal would start, punctuated by muffled remarks like: "Wait until our turn comes; we'll show you what good cooking should be!" Then deep silence. When the board was clear of all semblance of food, then, and then only, would the conversation begin.

The sea, upon which we had counted for fresh fish, rarely lived up to expectations. At six o'clock one morning we pulled in a thirty-pound dolphin on our troll. Tehate fixed it up for cooking. Spirits, which had become gloomy from monotonous diet, rose. Afternoon found us despondent again. We had done away with the fish. No fish . . . no fish . . . no fish . . . was written upon each hungry face. At five o'clock we hauled another in: thirty-five pounds! At breakfast next morning the cooks were heard murmuring: "Just two pounds left! God! We'll starve on this hell ship!"

Cookbooks were usually perused behind the pages of one's private log, as we read Nick Carter in school. Suddenly, something appeals. Midway between meals the usual mealtime row is heard in the galley. Pans crash. Flour flies out of the open scuttle . . . a white stratus falls away to leeward. Hours pass. The bell rings. The crew descend. They flop on the benches like an Albemarle thunderclap. They bang the table with their utensils.

The inspired cook, with ever so small a smile about his face, carries in a pie.

"Take a look at this!" he cries.

And faces fall.

"Is that all?" they chorus.

The skipper frowns:

"Is *that* the mean output of your cyclonic efforts?" he asks.

"Well," the cook tries hard to explain, "it's this way—"

But we all know the symptoms. He let the book get him.

Once we had pie and nothing else in the North Atlantic with all hell letting loose outside. We had chocolate pudding in a squall that all but blew the ringbolts out of the deck. We had a queer butterscotch concoction that shivered when the navigators were trying to stare down the haze for the Manzanilla landfall at Panama. We lived for days in horror of and in the solemn wish that no more spaghetti would come our way. One cook had a spaghetti jag. We even had it for breakfast once when he was awakened out of a sound sleep to cook.

One day a conscientious cooking crew read in their book the following: "This recipe serves six people with leftovers for Rover." *Director* had six souls aboard. Since we had a parrot, a young snake, and a very sizable Rover, the crew added a pound here and a quart there to the original recipe and rang the old Brion when ready to serve. Result: The serving plate never got out of the galley and the cooks looked rather ill-fed. Rover, José, and the boa and the chickens all showed up for a meal they didn't get. The crew supped on a seven-pound tin of New Zealand jam and an armful of hardtack. We had learned our lesson. The next meal was the total of all the ingredients called for, for the dish, in five cookbooks. This time nary a grumble! Needless to say, reefing points on trousers were soon obsolete on *Director*.

One day we heard the stores keeper muttering some unintelligible numbers. Upon being questioned, he said: "Four and a half tons," and went on counting his fingers. Pressed further, he explained: "Four and a half tons of food in four months! That's nine thousand pounds in one hundred and twenty days which makes seventy-five pounds a day, or 12.5 pounds per man per day plus bananas and oranges, papayas, lemons, limes, fresh meat, and other little items like rum swizzles, and gin gaspers." And we had had the idea that we could finance it for three years.

·　　·　　·

The clock-work changes of the night watches are finished. The stars have paled and disappeared. A ship's day is beginning. The sun reaches up, catches the hard rim of the sea, and laboriously chins itself over the edge.

The helmsman's back warms. He stops smoking, sheds the blanket from his shoulders and waits for the smooth that will let him run forward, leaving the wheel alone, to call Hey Hey. Two big ones lift her and she scuds forward happily. He leaves the wheel and runs down the starboard deck—the dory being in the way makes the port passage too slow—reaches the hatch, shoves it back and shouts down: "Hey Hey! Coffee!"

Then a sprint back to the waiting wheel. She is near the point of jibing, or coming round into the wind, but she never goes too far. A little wheeling and she settles back into her regular swing. Yes, Hey Hey's up and about. You can hear him pumping up the stove. You hear the merry roar of the flame. The coffee pot is being filled. You hear it plunk upon the stove. Now Hey Hey will relieve you. He comes up through the hatch—red pareu wrapped around his middle—stretches his enormous black arms, looks sleepily around the horizon, searches from bow to stern for flyingfish that may have flopped into the scuppers during the night—breakfast fish.

Bruce comes on deck and looks about, towel around his waist.

"How'd she do in your watch?" he asks.

"About five," you say. "What's your guess on the run?"

"Oh, about a hundred and fifty!"

"Optimist!"

Hey Hey relieves the wheel. It is almost seven now and the day watches will begin. You won't have to steer again until late afternoon. You've had the best watch, the one everybody gets every three or four days as the watches advance. You were able to sleep until everyone else had steered during the night. You got a lot of sleep. Now you'll have coffee.

You pick up the matches that have dropped to the deck in the night. You pass the galley ventilator and smell coffee, then you drop below into the fo'c's'le. Breakfast consists of cakes—like those in Childs—covered with honey. Gallons of hair-curling coffee. A general cleanup takes place after breakfast. Tobacco is swept off the chart table. Blankets are folded and put away. The heat is much too great for more than a short trick at the wheel. We take turns. Each must total three hours before seven in the evening.

At eight o'clock Denny goes to the deck with a sextant for the morning sight. Sheridan remains below to read the chronometer. By this time José has finished breakfast too, and is busy cleaning her beak on the mahogany trim around the hatch.

"Stand by!" Dennis shouts.

José yells.

"Mark!" says Dennis.

And José yells again.

"Right!" says Sher as he puts down the time at Greenwich.

"Twenty-one de-grees, eighteen min-utes and thu-ty sec-onds!" shouts Dennis above José's raucous screams.

"Right!"

That is repeated three times with José becoming more excited each time and more convinced that it is all done for her benefit. Then the sights are worked out.

By ten o'clock the heat has become so intense frequent showers of salt water are necessary. One man scoops up the water with a bucket from the sea, and dashes it over the other until he himself is pouring with sweat from the effort. Then he takes his turn.

Bruce decides to overhaul the engine. All the tools are dragged out and great noises come up through the engine-room ventilator. As noon draws near he quits the glory hole to make a batch of crumbly pastry for lunch.

José arrives late for lunch, flying down from aloft with a loud volley of curses to make a three-point landing in a bowl of hot soup. She burns her feet.

The day's run is announced at lunch and the usual post-mortem is held. If only that wind had held through the middle and morning watches we might have broken two hundred. Tiger wakes up aft and realizes lunch is on. He picks his way gingerly forward to a point where he effectively blocks the passage and turns his huge liquid brown eyes up at us—we feel as though we have denied him his right to live.

The afternoon watches are easy and soon it is six o'clock. At six the sun goes down—it always does in these latitudes, just as it always comes up at six. Supper over, the entire crew pours from below as though they're rats from a hole and breathe deep the first air of the day that doesn't seem like a blast furnace exhaust. We all bring huge mugs of coffee with us, and a pipe. Mattresses are hauled out and put on the cabin tops. Conversation is either reminiscent or controversial. We all speak with the complete authority of having read everything aboard on the subject. Then—and not as a slight to the company—someone doesn't answer when spoken to. He's asleep.

The Trade clouds march in their regimental regularity across the star-flecked sky. The helmsman leans back and steers with his feet— keeping Venus, there, off the port fore swifters. Odd things happen on these long night watches. In the daylight, of course, they seem silly, but when Venus starts her feminine flight across the heavens each night, strange things, impossible things, are commonplace.

The ship runs easily on the steady Trade. The sails loom high above you. By instinct now you wheel her this way and that, keeping the point of the compass course on the little white line. Time, so important a dimension ashore, has ceased to exist and fantasy comes rushing. Looking up, the heavens become personal; stars seem to come close. Then you remember that the light of those that you see—blue, red, and green, winking—started perhaps a million years ago, started on their flash to earth; that that light, so constant now, may have ceased to exist centuries ago. With difficulty you drag yourself from the stars back to earth. You fly back through the light years in between—to the sea. There it is, reassuringly there below you. You see a tiny ship before you. Her sails are lit by the planets—the sea marches in long even lines that come and go over the edge of the earth, toward her, to her, and then pass. You come down, down, and settle there.

You wheel her this way and that, keeping the point of the compass course on the little white line. Your journey for the night is over. Your circulation has almost stopped. You get up and sway to the motion of the ship. Time is a dimension again. It seems as if two hours must have passed and you squint at the watch in the dim light of the binnacle. You find you have steered for forty-five minutes. A flyingfish hits the deck near at hand and you retrieve him and place him where he won't wash overboard. An hour passes. You look at the watch. Now certainly you have steered your full time . . . fifteen more minutes. You fold your blanket, light a last pipeful. You give the next man five minutes' grace and then wake him. He comes to relieve you.

"West by south, error south," you say.

"West by south, error south," he answers.

And as soon as his answer is out your watch is over, your trick of responsibility is done.

And that is life in *Director* day after day. Day after day and week after week, until finally it was to become year after year.

⚓

31

Forty Thousand Miles in a Canoe

CAPTAIN J. C. VOSS

(Of all world-girdling small boat voyages, none makes such ex-traordinary reading as "The Venturesome Voyages of Captain Voss." *A strange individual this skipper must have been, sailing his peculiarly rigged Indian dugout, the* "Tilikum." *Amazing were his adventures—as the following extract from his book will testify,—so much so that until he convinced the experts with actual evidence, few people believed him. —Ed.)*

In describing *Tilikum* Voss says:

"The canoe as I bought her was made out of one solid red cedar log; in other words, she was a proper "dugout," as used by the Indians for traveling about, propelled by means of paddles, and at times, when the wind and weather was in their favour, a small square sail was used. Red cedar is very durable, but soft and easily split. I was therefore obliged to take great precautions in strengthening her so that she would be able to withstand the rolling and tumbling about in hard sailing or probably in the heavy gales she would most likely encounter during our trip. To put the little vessel in seaworthy condition I bent, twenty-four inches apart, one-inch square oak frames inside the hull from one end to the other, fastened with galvanized iron nails, and as the canoe was not quite deep enough for my purpose I built up her sides seven inches. Inside of the vessel I fastened two-by-four-inch floor timbers, over which I placed a kelson of similar measurement, and fastened the same with bolts to a three-by-eight-inch keel. On the bottom of the keel I fastened three hundred pounds of lead. She was then decked over, and I built a five-by-eight-feet cabin in her and a cockpit for steering,

Reprinted by permission of Dodd, Mead & Co.

after which I rigged her with three small masts and four small fore and aft sails, spreading in all two hundred and thirty square feet of canvas. The masts were stayed with small wire, and all running gear led to the cockpit, from where the man at the helm could set or take in all sails.

"When all was ready I broke a bottle of wine over her figurehead and called her *Tilikum,* an Indian name the meaning of which is "Friend." The *Tilikum's* dimensions were then as follows:

Length over all, including figure-head .... 38 ft.
Length on bottom .... 30 ft.
Main breadth .... 5 ft. 6 in.
Breadth, water-line .... 4 ft. 6 in.
Breadth, bottom .... 3 ft. 6 in.

"I then put on board half a ton of ballast, which was placed between the floor timbers and securely boarded down, and four hundred pounds of sand in four bags was used as shifting ballast to keep the boat in good sailing trim. About one hundred gallons of fresh water in two galvanized iron tanks was placed under the cockpit. Three month's provisions, consisting mostly of tinned goods; one camera; two rifles; one double-barrelled shot gun; one revolver; ammunition; barometer and navigating instruments completed our equipment. With everything on board, including ourselves, she drew twenty-four inches aft and twenty-two inches forward."

While in the South Pacific, the *Tilikum* had on various occasions carried mail matter from one island to another, but along the coast of New Zealand the people made her a regular mail-boat. Nelson was no exception, and we had a considerable mail taken on board here, mostly for Europe. The *Tilikum* was almost ready for sea, when an elderly, well-dressed lady brought down a tin box of ample dimensions and asked me to take it along to the Cocos Keeling Islands in the Indian Ocean, having been informed that our course laid that way.

"Madame," I protested, "you will please pardon me, but the box will take up too much space in our small vessel." She then begged and pleaded with me, confiding that she had a son living on the islands who was employed in the British cable service, and that the box contained a fruit cake which was intended for a birthday present to him. My mate happened to be standing near by, and on hearing "fruit cake"

—later on he gave sufficient evidence that he liked fruit cakes quite as well as the ladies—whispered into my ear: "Take it, Captain, take it; by all means take it along;" and as the good lady at the same time was appealing so insistently on behalf of her beloved son, I finally agreed and promised to do my best to deliver the parcel. My acquiescence nearly brought the tears into her eyes, and to show her thankfulness, shortly before we sailed she sent on board another cake for the entertainment of the crew of the *Tilikum*. Buckridge said that the second cake was to pay freight and insurance on the first one, but also sarcastically remarked that if fruit cakes were as scarce in the Indian Ocean as they were at the South Pole, something might happen to the mail before we reached the Cocos Keeling Islands! "Look here, Buck, broaching cargo is prohibited on board the *Tilikum*."

Many mornings later we had lost sight of the New Hebrides and were sailing over the Coral Sea towards Rain Island Passage in the Great Barrier Reef. The Coral Sea is considered the most dangerous portion of the Pacific to navigation, owing to numerous small islands, shoals, reefs, and the strong currents prevailing there. However, by carefully maneuvring our boat and keeping a sharp lookout we got across without a hitch and safely arrived at the "Barrier." This great reef is unequalled for its vast extent, and is a formidable obstruction to sea traffic. It runs for about a thousand miles along the north-east coast of the Australian continent and extends still farther to New Guinea. At low tide the reef is nearly level with the water, and numerous small black coral rocks appear here and there in the heavy surf, which ceaselessly breaks all along the great barrier. It is so low that it can be seen only within a distance of about four miles from a vessel's deck, and owing to the strong current setting through from the east, the passage is extremely dangerous after dark. Rain Island Passage forms one of several gaps which ships going east or west may use. The width of the channel is about five miles, and Rain Island, which is marked by a beacon, is almost in the centre.

Shortly after clearing the passage we sighted a schooner lying at anchor near a small low island. The wind being light and the water smooth we sailed alongside, and on invitation from the captain went on board. The latter was a Japanese from Thursday Island, the schooner's home port, and was engaged in fishing for "bêche de mer," the edible sea-cucumber so highly valued as an article of diet by the Chinese.

We stopped there for about two hours, watching the divers and talking to the captain who was himself busy spreading and drying the slugs in the sun. I noticed several large sharks hovering round the vessel, and with regard to them asked the captain whether it was not dangerous for his men to dive whilst they were about. He assured me that he had never lost a man through the medium of a shark although he had followed this business for years. This statement is in accord with my own observations. The surrounding water was infested with sharks, but, as in many other localities, where there is every opportunity to feed on human flesh the so-called man-eater never even attempts to attack a man. No doubt there are species more ferocious than others, but I have come to the conclusion that the average shark is not so bad as his reputation.

At ten o'clock the *Tilikum* resumed her westward voyage under all sail and with a light breeze. During the following two days we accomplished a hundred miles of our course, and were then becalmed for three days. During this time we experienced the hottest weather on the whole cruise. If it had not been for the awning spread over the after-end of the boat in all probability we would have melted into grease spots.

The sky was clear during the three days of calm, and with the aid of my sea telescope we could see plainly to a considerable depth into the waters beneath us. It was here that I saw the greatest variety of sea life I ever have witnessed during my travels at sea. Some big, grey sharks lay motionless quite near our boat, while others could be seen moving lazily to and fro further down. Each of the monsters was accompanied by a herd of pilot fish. The latter are small and striped greyish black. About a dozen of them would busily ply round the head of a shark, sometimes entering his mouth, but soon came out again. On one occasion I noticed a shark, twenty feet in length, apparently asleep within five feet of the bottom of the boat. Through my telescope I could distinctly see that his eyes were shut; the only part of his long body that moved was the tail, which slowly swung from side to side. His pilots, however, seemed to be on the alert; they kept swimming around the big fellow as if on the lookout for danger.

Besides the sharks and their pilots there was a great variety of fish of different size and colour. These were either motionless or sluggishly moved about as if also affected by the heat of the day. I observed a yellow sea snake, from three to four feet in length, pass through the school of fish in a great hurry. There were many more of these venomous reptiles twisting and turning in all directions below or whipping along almost on the surface of the sea. But one and all, wherever they were bound, showed the same haste, in contradistinction to the finny philosophers near by. The appearance of these strange creatures was sufficient to give one the "creeps." With the exception of the mermaid and the mysterious sea serpent, this representation of marine life seemed to be complete.

Light variable winds, alternating with more calms and most beautiful weather, followed. These conditions lasted for nearly a month. One evening I was sitting in the cockpit, smoking my pipe and waiting for a breeze to take us into the Indian Ocean. Suddenly I felt a tremor through the boat similar to an earthquake.

"What is that?" Mac shouted out of the cabin where he lay in his bunk.

"An earthquake," I volunteered, wondering myself what it could be.

"It must be a seaquake then," my mate responded.

Just then we received another shock and it dawned upon me that some large fish must be rubbing or bumping himself against the *Tilikum's* bottom, which caused her to shiver from end to end; so immediately I seized an oar and gave our unwelcome visitor, whoever it may have been, a good and hard dig in the ribs which induced him to make a speedy retreat.

Near Bathurst Island, on the south side of the Arafura Sea, is an expanse of shallow water called the Mermaid Shoals. I had been told by seamen that the famous mermaids there had their abode. Others connect the name with the simple report that a ship of that name had been wrecked on the shoals. Be it as it may, we decided to visit the place, and altered our course to that direction.

On approaching the home of the sea beauties we parted our hair in the middle and made ourselves look respectable, as if going to a Sunday afternoon church meeting. When we reached the shoals the wind became light and the water smooth. So we again brought our water telescope in position and kept a sharp lookout for the ladies. We saw many fish loafing about, but whether they were "mashers" or not it was difficult to conclude. As in other places, these waters teemed with life, but the mermaids were evidently out on a picnic, for we could not discover their whereabouts. Of course, it was the old story; my usual luck with the ladies!

In such manner we had spent a month on the Arafura Sea, drifting about and only occasionally helped by light winds which gradually transferred us into the Indian Ocean. Then the south-easterly trade winds took us in charge and made the *Tilikum* again spin along at her best. The steady cool breeze came as a great delight after the hot time we had experienced in the tropical Arafura Sea, and we made from a hundred and thirty to a hundred and fifty miles daily towards the west. In the morning of November 8th we sighted the Cocos Keeling Islands. When within four miles of the south-eastern end of the group the wind died out and once again we were becalmed.

"Just as the doctor ordered it," I said. "Now we have a few hours in which to clean things up and get dressed, and when the breeze comes in this afternoon we shall sail into port in style and deliver our mail!"

But sailing vessels will always be victims to uncertainty, and no one can rely upon the time of their arrival. So it was in our case; I had made a mistake in the calculation. If we had known what lay before us we could easily have pulled in to land, but we did not, expecting the wind to come up every minute. However, we drifted all afternoon, and to kill time put everything on board in order. The following morning we found to our great surprise and consternation that a strong current had carried us nearly out of sight of the islands. During the forenoon a light breeze commenced blowing from the east and we tried to beat back to Cocos Keeling. But despite all our efforts the current forced us to the west and the islands finally disappeared below the horizon. It was not only for the promise I had made the old lady at Nelson to faithfully deliver the parcel to her son, and which could not now be fulfilled, but still worse, we had only eight gallons of fresh water left in our tanks, and the next land on our course was Rodriguez, a small island about two thousand miles to the west-south-west of the Cocos Keelings. Calling to mind the good old lady who brought the cake on board I felt sure that she would forgive me under the circumstances, but to start a two-thousand-mile trip across the ocean with only four gallons of water per head aboard was somewhat disheartening!

The weather was fine, but the light easterly breeze and current setting to the west made it absolutely impossible for us to return to the islands. I was fully aware of the fact that whatever happens on a sea voyage, be the vessel large or small, the master is responsible. And accidents will happen sometimes, but nine out of ten so-called accidents are due to carelessness or ignorance. When we sailed from Thursday Island we had our tanks full, enough water to last us for seventy days. We were now forty-three days out and having only eight gallons left I could not help accusing myself of carelessness, and the weight of the thought lay heavy on my mind. The *Tilikum* was a poor vessel to sail by the wind, and to take a course for Rodriguez with the light breeze then prevailing and eight gallons of water in the tank meant almost certain death from thirst. However, in going on we had two chances in our favour. One, that we could be benefited by a strong Trade wind which would enable us to make the run in about sixteen days, and by being economical the water supply would just last for the voyage. The second chance was the possibility of our meeting with rain on the passage. Consequently we decided to reduce our daily allowance to one pint of water each, and set off for the west.

One pint of water per day is very little in warm weather, but circumstances compelled us to follow the lines we had laid down and use the little we had in a way that would give the best satisfaction for quenching our thirst. To begin with, we ate all our food cooled and without adding salt. Potatoes were boiled in salt water and eaten after cooling. Instead of consuming our pint neat, which would have been an unsatisfactory and therefore uneconomical way, we mixed half with a little oatmeal and used the remainder in making coffee. Both beverages proved to be greater thirst-quenchers, and we drank sparingly. We also took frequent salt-water baths, and kept the deck and cockpit wet. Neither of us was in the habit of chewing tobacco, but at times, when our daily allowance was exhausted, and we felt dry, we found that great relief was obtained by keeping a small piece of tobacco in the mouth. By adhering to these rules we trusted that we would be able to keep alive for from sixteen to twenty days, and in such condition as to be capable of sailing our boat to Rodriguez.

The first and second days passed, and on the third the weather was still warm and the wind light. Mac said he would give ten pounds for a good drink of water. I would have given that myself, as water under the circumstances was much more precious than money. And if the *Tilikum* had been loaded down with gold and diamonds it would not have made an iota of difference: we had to get along as best we could on our allowance!

On November 13th four days had passed since we kept off to the west. During this time we ran off two hundred miles of our course; indeed very slow progress. There were no signs of rain or of the light wind improving, so we did what people in Australia do during a long drought—prayed for rain. It seems strange, but nevertheless true, the same night the breeze increased rapidly and the sky became cloudy. On the following morning we had all the wind our boat could stand under all sail, and to make things still more hopeful heavy clouds darkened the sky in the south-east, a sure indication of approaching rain. Mac opined that our prayers were about to be answered. At noon the clouds hung over our head, but as so far there was no rain our midday meal consisted of cold unsalted food and a little cold coffee, as before. Meanwhile the heavy clouds accumulated, and an hour later it began sprinkling. Not heeding the favourable and strong wind which sent the *Tilikum* along at her best, we lowered all sail and spread them over the deck to catch the precious fluid. We had hardly done this when the rain came down in torrents. While this lasted we employed ourselves with buckets, and by three o'clock our water tanks were filled to the top!

The wind kept fresh and fair, but with our tanks full we were not in a hurry to get under sail again. First of all, we lit the stove to prepare a good drink. While thus engaged, I said to my mate: "I say, Mac, try to reef yourself through the after-hatch, and right at the end of the boat you will find a square tin box; fetch it up and bring it into the cabin." Mac, no doubt, had a faint idea that there was something good in the box, for without a word he disappeared, and in a few minutes I saw the box coming through the hatch and Mac's face behind it.

"What's in this?" he enquired, on entering the cabin.

"Open it and see," I replied, and he certainly lost no time. There before us was a fruit cake, to all appearance as good and as new as on the day it was made. Mac looked at it in astonishment and believed it was God-sent. "Not exactly that, Mac," I assured him, "the rain was God-sent, perhaps, but this cake was sent on board by a kind old lady of Nelson, New Zealand, to be taken to her son who is in the British cable service on the Cocos Keeling Islands; therefore, it is not my property; but under the circumstances, what do you think we ought to do?"

The coffee was ready by that time, and Mac asked me to pass him a knife and he would tell me what he thought of the New Zealand lady and her cake; and eventually we were agreed that it was one of the very best we had ever tasted, and as we had been without such luxuries for so long a time, and being now provided with plenty of water, I am sure neither Mac nor myself will ever forget how we appreciated the coffee and cake aboard our little vessel that afternoon on the bosom of the great Indian Ocean. And I sincerely hope, as there had been no possibility of my delivering the parcel, that the good lady and mother, on receiving my letter of explanation from Durban, South Africa, has forgiven me!

The wind kept up its force, the rain continued pouring down, and after finishing our coffee we enjoyed a smoke and a sociable chat, congratulating ourselves on our good luck. Then we hoisted sail and with the strong trade wind—the heavy rain preventing the seas from breaking—the *Tilikum* went flying to the westward again.

The rain had been the most welcome occurrence that could befall us, and we therefore did not register any complaints when its ceaseless downpour made things rather uncomfortable for us. We simply kept a-going and looked pleasant. It rained all the afternoon, night and next day, and when it did not stop on the third day we thought that another great flood was descending, and we might be doomed to sail for ever. The wind stirred up a small gale, and only the heavy rain, by keeping the high seas from breaking, enabled us to keep our course. These conditions lasted a whole week, for eight days, during which time we saw neither sun, moon, nor stars, and we were thankful when the rain ceased and the weather cleared up. Then the wind died out and we found ourselves becalmed once more in the nicest weather imaginable. For the first time since the rain had begun we took our position and ascertained that we had made a little over twelve hundred miles during the deluge.

Calms, as a rule, are considered unwelcome visitors to sailing vessels, but as almost everything on board had become wet we appreciated the change of weather which afforded us an opportunity to dry things. On that warm sunny day we lowered all sail and hung our wet garments about the deck and rigging, which made the *Tilikum* look more like a floating Chinese laundry than a sailing vessel. Thereupon we spread an awning over the cockpit and, devoting our energies to a comfortable smoke, took the world easy.

It was shortly after sunset, and the beautiful tropical stars twinkled overhead. Dolphins were indolently swimming around the boat and now and then one would roll over on its side as if to turn in for the night. While sitting in the cockpit I noticed a fat fellow in that attitude within easy reach and, thinking that I could master him, I grabbed his tail to pull him aboard. In this, however, I was much mistaken. No sooner did I touch him than he dealt me such a blow with his tail that it reminded me of an electric shock. I was only too willing to let go, and off he went like a shot. About eight o'clock, when I had retired to the cabin and Mac was still flirting with a dolphin, I suddenly heard heavy splashes followed by a tremendous noise in the cockpit. My mate sang out, "What is this? What is this? I have caught the very devil himself!" Out on deck I went like a flash, and found Mac scared to the top of the cabin deck, keeping well out of the way of something struggling in the cockpit, nearly smashing in the floor in an attempt to escape. Owing to the darkness I was unable to identify the noisy stranger, but on getting a light I quickly recognised a young shark, about three feet and a half in length, making desperate use of his tail. In accordance with seamen's etiquette the world over we despatched him and threw the body overboard.

We were now within four hundred miles of Rodriguez. The south-easter kept blowing fresh, and on the morning of November 28th we sighted the island. In the afternoon, when approaching the southeastern end, seeing quite a number of fishing craft going in towards the land, we followed them. By and by we entered the coral reefs by which part of Rodriguez is surrounded and soon dropped anchor among the fishing boats which had arrived just beforehand.

Having finished our meal we were invited ashore by the fishermen, who numbered about forty in all. They had a camp, consisting of a few roughly-built houses along the beach, where they slept whenever time did not admit of the walk to their homes which were some distance away in the town on the northern end of the island. To the latter place we sailed the following morning, and on arrival made fast to a small wharf. A little later the magistrate of Rodriguez came aboard to acquaint himself with the particulars of our voyage. On invitation I accompanied him to his house, Mac staying aboard, and with him and his family spent a pleasant day. In the course of the afternoon, while we were sitting on the spacious veranda talking on various subjects, a gentleman stepped up to me and said: "Are you Captain Voss of the *Tilikum?*" I affirmed the question, when the stranger introduced himself as the manager of the British cable station, at the same time handing me a cablegram which read:

"TELL CAPTAIN VOSS TO SEND THAT CAKE."

"What is the meaning of this, sir?" I enquired. The manager explained that he had been on board the *Tilikum,* where my mate, amongst other things, had recited him the history of the New Zealand fruit cake and its unforeseen though glorious end. In possession of this news he had at once cabled to his colleague on the Cocos Keeling Islands, with the result that the indignant consignee sent back the above stern demand. The gentleman evidently found it difficult to suppress a smile when adding that he had further received instructions to immediately take steps for recovering the cake.

"Now, what are you going to do about it?" he asked.

To reproduce a cake which had been consumed head and tail a fortnight previously, was certainly impossible. So I gave a short explanation why we did not reach the Cocos Keeling Islands despite all our efforts, and of the troubles encountered in the Indian Ocean—not forgetting, in conclusion, to praise the high quality of the subject under discussion—and finally I asked the manager to express my regret to the consignee and transmit him my apology. Both gentlemen had a hearty laugh, and we all agreed that the matter was settled.

A short time previous to our arrival a cyclone had visited the island, leaving considerable devastation in its track. My host said it had been the worst on record on Rodriguez. A flagstaff had been erected near the magistrate's residence, made of hard wood, nine inches in diameter and measuring thirty feet from the ground to the cross-trees where it was stayed with four strong wires. But the force of the hurricane had snapped it at the middle like a piece of glass. On my enquiring about the velocity of the wind the manager of the cable station informed me that he had no record, as his anemometer, together with the building in which it was mounted near the station, had been blown to pieces and scattered so thoroughly that he was unable afterwards to recover any of the fragments. Iron pillars that supported the roof of the veranda at the cable station were broken and houses razed to the ground; much damage had been done all over the island. "If the cyclone had struck you in your little vessel," the manager added, "you would never have lived to tell the tale!" "I am not so sure about that," I replied. "I think our risks on the open sea would not have been any greater than yours on the island here, with broken flagstaff, pieces of iron pillars, houses, trees and other heavy material flying about your ears."

Thereafter, we set sail for Durban.

⚓

32

Nathaniel Bowditch

COMMODORE E. S. CLARK, JR.

THE name "Bowditch" is familiar to all naval, Coast Guard, and Merchant Marine officers, as well as to all other men who are connected with the sea, but we usually think of a book published by our Hydrographic Office rather than of that fine, brilliant, and lovable character who was the author of the original edition of the *American Practical Navigator*.

Nathaniel Bowditch was born on March 26, 1773, in the town of Salem, Massachusetts, the son of Habakkuk Bowditch, a poor cooper who later turned to the sea as his business ashore dwindled. Young Nathaniel was in school but a few years when his father grew even poorer, and at the age of ten years and three months he was obliged to leave school to serve an apprenticeship in a firm of ship chandlers by the name of Ropes and Hodges. Even at this age he was much interested in mathematics and availed himself of every opportunity to study and read books on the subject. At work, every spare moment found him in some corner studying. He read everything that he could secure on the works of Sir Isaac Newton and at this young age started to study Latin. Judge Ropes of Salem had quite an extensive library which he put at the disposal of the boy, and at the age of fourteen Bowditch was making a special study of navigation and astronomy. During this year he made several crude instruments and actually wrote an almanac which remained in his library as long as he lived, and is still in the Bowditch Library in Boston.

Between the ages of fourteen and eighteen he studied Latin and algebra very seriously. At nineteen he studied French from a native Frenchman, offering to teach English in exchange for French. About this time the firm of Ropes and Hodges went out of business and young Nathaniel was obliged to seek employment elsewhere, going to work for Samuel C. Ward, another ship chandler with whom he remained for two years.

Late in the summer of 1794 he had the desire to go to sea and signed on board a Salem vessel as clerk. During this voyage he availed himself of every opportunity to study navigation and to make as many observations as possible of heavenly bodies and their movements. On January 11, 1796, the *Henry* dropped anchor in Salem Harbor having completed her voyage of just one year.

Bowditch was ashore but a short time when he started out on his second voyage. As before, he spent much time in studying and gathering all possible information relative to the subjects in which he was so much interested.

While at Cadiz some years later he visited the observatory with Count Mallevante, who likewise was much interested in astronomy, and Bowditch agreed upon returning to America to send the Count several publications on the subject which he much desired. On his other trips and while on shore he had made quite a study of a number of volumes published on navigation, among them one in particular which was much in use at the time, a volume published by a London firm and written by Hamilton Moore. In studying it carefully he discovered many mistakes in its tables (some 8,000 in all) and many misstatements which he believed to have been to blame for many shipwrecks. This book had been republished in America in 1798 by Mr. Edmund Blunt of Newburyport, Massachusetts, who was about to get out a second edition but hearing that Mr. Bowditch had found much wrong with the present edition he asked to confer with him. Mr. Blunt finally prevailed upon Bowditch to take several copies of Moore's *Navigator* along with him to study and upon his return he would get out a new third edition, containing all of Bowditch's corrections. While at sea Bowditch was busily at work checking tables and statements and found so many of them in error that he decided it was useless to try to correct them all and made up his mind to write a complete new book of his own. Much of his spare time was now spent on this work. Upon arriving home from this voyage he was determined again that he would now remain ashore and settle down.

For two years after his last voyage Bowditch was a merchant in

Salem. During this time ashore he was appointed to many positions of trust in the community, including membership on the school committee of Salem and the secretaryship of the East India Marine Society, the purpose of which was to assist in the relief of families of deceased seamen and the further promotion of the art of navigation. To be eligible for membership one must have been around Cape Horn or the Cape of Good Hope as a master or supercargo. Bowditch became a very active member of this society and contributed much to its cause and work.

He was now putting in more work on his book and conferred from time to time with Mr. Blunt, with the result that in 1802 instead of there being a third edition of Moore's *Navigator* published, Bowditch came out with the first edition of his *American Practical Navigator,* published under his own name with Mr. Blunt as the proprietor. Bowditch was 29 years of age at the time. This volume soon replaced Moore's book in the United States and in a short time found its way to England and other foreign countries.

At the time of its publication the East India Marine Society of Salem appointed a committee to thoroughly investigate its contents and the committee submitted a report recommending its use to ship captains and seamen. The original manuscript of this report is now in the Peabody Marine Museum of Salem, along with the original manuscript tables computed by Bowditch and used in this first edition.

Before his death in 1838 he had lived to see ten editions of his book come off the press, representing over 30,000 copies. Some thirty odd editions were published before the government took it over and published it through the Hydrographic Office. The Peabody Museum has in its possession copies from eighteen of the early editions.

During September of 1802 Bowditch with three other men from Salem purchased the ship *Putnam* recently built in Danvers, and on November 21st she sailed with a valuable cargo under the command of Nathaniel Bowditch himself, this time sailing from Beverly and heading directly to the Indian Ocean by way of the Cape of Good Hope. On this voyage Bowditch assigned duties to his various officers in such a way as to leave himself much free time in which to complete several works already started by him. His outstanding piece of work on this voyage was the translating from French into English of the *Mécanique Céleste* by the French mathematician La Place.

With the publishing of his *Practical Navigator* and the popularity he was gaining on shore, honors were being bestowed upon him. In 1802 he was honored by Cambridge College, in 1803 he was offered a professorship in mathematics at Harvard, in 1804 he was installed as President of the Essex Fire and Marine Insurance Company, in 1810 President Jefferson asked him to accept the office of Professor of Mathematics at the University of Virginia, and in 1820 the Secretary of War asked him to serve as professor at the Military Academy at West Point. All of these professorships he refused because he disliked talking in public. He delighted in offering instruction to worthy young men in private but often declined to appear in public.

Between the years 1800–20 he wrote 23 papers that were published in the *Memoirs of the Academy of American Arts and Science* relating to observations of the moon, comets of 1807–11, eclipses of the sun 1806–11, measurements of the height of the White Mountains in New Hampshire, observations of the compass, a pendulum supported by two points, and corrections in Newton's *Principia*.

In 1818 Bowditch was chosen a member of the Royal Society of London and Edinburgh and was enrolled on the list at the Royal Irish Academy. He was made an associate of the Astronomical Society of London, the Academies of Berlin and Palermo and at the time of his death he was being chosen a member of the National Institute of France, which had only eight foreign members at the time. In 1816 the degree of LL.D. was conferred upon him by Harvard University, after which he was always referred to as Dr. Bowditch. In 1829 he was appointed President of the American Academy of Arts and Science of which he had been a member since 1799.

In 1838 he became seriously ill and was confined to his home where most of his last days were spent in his library and with his children. Death took him on March 16, 1838. His passing was mourned by all in Boston and the Boston Marine Society at the time resolved:

"As an astronomer, a mathematician, and a navigator himself, a friend and benefactor has he been to the navigator and seaman, and few can so justly appreciate the excellence and utility of his labours, as the members of this Society. . . . His intuitive mind sought and amassed knowledge, to impart it to the world in more easy forms."

We who today are following the sea in one form or another nearly one hundred years after his death are still using these "easy forms" given to us by this "friend and benefactor to the navigator and seaman."

33

Force 12, Beaufort

ALEXANDER CROSBY BROWN

As THE *Chance* approached closer, Bermuda, sparkling like an emerald in the sunshine, rose out of the sea. A line of frothy breakers surrounded the shore and, looking down through the clear water around the ship, we could see dark, fantastic coral shapes on the bottom, ten fathoms below. Once through the narrows, the entrance broadened into the azure lagoon. We hove to for the port doctor, and then, since the wind had fallen, it was decided to take in sail and taxi the remaining seventeen miles around to Hamilton Harbor, in order to arrive before nightfall. The water was quiet now, and deep blue tongues stretched far inland, where they met green rolling hills. Dazzling white-roofed houses dotted the landscape.

On we went, the channel twisting now as the ship made a long bend away from the friendly shore. I hesitated to go below, lest this miracle, land, be spirited away entirely. Then the route curved back and the *Chance* entered a long bay. The village of Hamilton was crowded on the northern shore—a church spire, myriad houses, a palatial hotel, docks with a large white steamer moored alongside.

"Let go!"

The chain rattled through the hawsepipe. We had arrived. In spite of the doldrums, the *Chance* had averaged one hundred and twenty-five miles a day for the long passage of 2,875 miles from Madeira. Now three oceans had been crossed, a mere six hundred odd miles remained and the voyage would be completed.

Despite the call of home, we lingered on in Bermuda, apparently for no reason. Perhaps this was because no one is anxious to leave a beautiful tropical island to return to a cold, gray New England winter. After a week had slipped by, however, we felt compelled to move. Accordingly, on September eighth the anchor was hove and the ship brought alongside Darrell's wharf to take on water and stores. The following day was dark and squally, but, nothing daunted, we hoisted the long "Homeward-bound" pennant and cast off.

Squall after squall obscured the channel markers, but ultimately we passed the last narrows' beacon and headed out to the open sea while darkness took possession of the sky. The last lap had started. Although we did not expect to duplicate the run down to Bermuda two years before, which had been made in only four days, nevertheless, we counted on making a reasonably good passage to the New England coast. Southwest winds prevailed and northeasters were common. Either meant a beam wind home. Moreover, September is late for West Indian hurricanes, which generally avoid the region in which we would be sailing by swinging out southward of Bermuda to dissipate themselves in mid-Atlantic.

Although we had delayed in Hamilton, when away from the land, all sail was set and we crowded her along full and by to a fresh westerly. The lighthouse on St. David's Head disappeared that evening, and when dawn broke Bermuda was no longer visible. The wind petered out and calms persisted throughout the day, until a Gulf Stream squall from northeast brought the ship a favorable, if somewhat damp, slant of wind. The mainsail, which had been lowered for the calm, was set again and we sailed on with easy sheets. This nor'easter continued with varying strength for two more days. Occasionally the sky cleared and we would be rewarded with a glimpse of the sun, but for the most part the weather remained overcast and squally. Since the Gulf Stream is notorious for its squalls, beyond bemoaning the discomforts of rain trickling down the neck of one's slicker jacket, we did not give them much thought.

The northeast wind was succeeded by more moderate weather and we even carried the fisherman staysail as the light breeze shifted more into the eastward. The sun graciously peeped out around noon, and, with a good sight, Skip fixed our position at three hundred and ninety miles southeast of New Bedford. That evening the wind had shifted more to the southward and the ship was running before it at a good five-knot clip. At 8:00 p.m. the barometer was high—30:25. Later, as I went below, I thought, "Well, if this keeps up . . ."

Perhaps a revengeful Fate had heard me.

Apparently there was something unusual in the atmosphere. It seemed extraordinarily oppressive and no one slept well. At 2:30 a.m. I came up on deck to discover two steamers, which were the subject of muffled comment of the watch. A squall arrived shortly after and their lights were immediately snuffed out. Leaving the watch to its uncomfortable task of steering in the rain, I went below. At four o'clock the wheel was changed and the watch noted in the log: "Wind, SE Force No. 3, freshening. Swell heavy, increasing. Rain and overcast. Log, 75¾. Barometer, 30:14." It looked as though the *Chance* were in for some dirty weather, for in four hours the glass had dropped six hundredths of an inch. In the sky gray light gradually substituted for black. No actual dawn occurred but for a moment the eastern horizon suffused with a feverish red blotch before returning to its dismal half tones.

"Red in the morning, sailors take warning!"

No need for this additional omen now. The wind was increasing and, although still running before it on the course, it was just a question of time now before the ship would have to be reefed down. In spite of the ill portents, we hated the thought of wasting that fair wind. At eight o'clock the glass had dropped to 30:08. All hands were below at breakfast and, although the helmsman cast many an anxious glance over his shoulder at waves dashing up astern, he judged that the ship could hold out a little while longer under her present canvas. Then suddenly a dark squall struck. Alone on deck, he rang the brass bell by the binnacle violently. All hands came tumbling up the forehatch, but no explanations were necessary. During a lull and between squalls the helm was put hard down and the bow swung around.

"Coming down on the peak. Slack the throat!"

Fighting and jerking, the mainsail came in, as we strained on sheet and downhauls. Heavy spray shot over the forward deck and the wet and twisted lines ran stubbornly through the blocks. With difficulty, for the deck was dancing madly, a line was passed through the reef cringle and the reef points tied. The helmsman squared off before the wind once more, but, now that the reef was set in the sail, it seemed foolish to hoist it. For with foresail alone, the ship had all the breeze she needed. Running before the wind is apt to be deceptive, for, by going with them, both wind and sea seem less. It was blowing hard and the contrast from running free to heading into the wind was quite apparent when we came up to lower the mainsail. There comes a time, however—sailing before the wind—when it is too late to swing around; for in this maneuver the boat turns broadside, and in that helpless condition a wave might break aboard that would send her to the bottom. To all of us it was obvious the weather was in no way moderating. With the long overhanging counter of the *Chance* we could not risk being forced to run before the storm. We decided to heave to while there was yet time on the starboard tack.

Again the helm was put down, but on this occasion we were not as lucky as before. A playful graybeard smashed its way aboard, filling the waterways and completely soaking everyone ,on deck. In another second, though, the bow swung around and the ship was hove to and riding the swells more easily. Still, the elements were by no means finished with us, for, as mounting waves hurled themselves toward the ship, we noticed the tops were rushing forward. In a little while that would mean a breaking sea. We would have to shorten down some more. The foresail could be reefed; surely that tiny patch of canvas could ride out the storm! It was unfortunate that, on the way down, the foresail jammed for a second, and, catching it when the peak was slack, a playful gust ripped out one of the seams. There was no time for repairs. *Hors de combat,* we lashed the sail as best we might and broke out the storm trysail. This, when set on the mainmast, made a balanced heaving-to rig with the forestaysail. Then for the first time we could sit down and take stock of our position.

It was not an enviable one. An hour and a half before we had been running under full sail. Now the ship was shortened down as far as she would go. There remained only the sea anchor. In the short intervening time the barometer had fallen another eleven points and was still on the downward path. We had expected that we might run into a northeast gale, but this monster was coming from southeast. All things considered, it could mean only one thing. And yet we hesitated to name it. There was nothing to be done at present. Everything was secure and the ship was riding as easily as could be expected. So, leaving a man on deck, all hands adjourned to the fo'c's'le for a game of poker.

It was a strangely silent game. Six of us crowded about the long table, a woolen blanket over it to prevent the cards from sliding, our legs braced underneath. There was a lifelessness in the rolling and pitching, although the ship lurched wildly. Occasionally a resounding smack aft would tell of a big one coming up under the stern, and you could almost feel an involuntary wince pass through the group. Above, the wind kept up a steady scream, ropes beat a tattoo on the masts, and spray hissed across the decks.

It was an odd time to be playing poker. All hands realized it and we knew that we played only to steer our minds away from the rising hurricane. Conversation was nonexistent; we played hastily, seriously, with furrowed foreheads and vaguely staring eyes, as if fortunes were at stake. I shall never forget the suddenness with which everyone tossed down his hand. Nothing had happened, no one said anything. But as if a silent electric message had flashed around, each made the motion at the same instant. We looked somewhat sheepishly at each other and straggled on deck. I remember a slight sensation of nausea that I wished I could have explained as seasickness.

A fine drizzle swept down from leaden skies and, with recurring frequency, low-hung squalls came shrieking down, pelting the sea with torrential rain. Awe inspirings, those squalls; like massive pyramids, they towered up to the heavens. Their bases, close to the raging ocean, were torn and ragged. Relentlessly they came, blinding, tearing, howling, obliterating all before them. The sea was breaking in long black ridges flecked with hoary crests. They, too, buffeted the poor ship and sent sheets of solid water across her decks. A full gale was raging. Spellbound, we watched the barometer's headlong descent: 29:98, 29:80, 29:75. "It's a hoodoo, let's t'row it overboard!" said Sandy.

Noon came, and now the lulls between the squalls were only of a few moments' duration. A momentary break, and then another, blacker than the rest, would encompass the ship. So this was a hurricane! All the way around the world we had successfully dodged them. Now, on the last lap of the voyage . . . Faced by the actuality, all we could do was say to ourselves apathetically, "Well, well! A hurricane. What do you think of that!" There was no answer. Our minds could not yet grasp the full significance.

Our last card was the sea anchor. If the ship would ride to that, we might get through. After a struggle with wet and kinky lines to make fast the heavy wooden crossbars, the canvas pyramid was rigged and weighted with a piece of pig-iron ballast. A brand-new length of manila cable, two inches in diameter, was attached to the bridle. One end of this led through the hawsepipe and made fast to the windlass. With feelings strangely mixed we cast it overboard on the weather side. The forestaysail might as well be taken down now before it blew to ribbons. Waist deep, Tom, Ed and I battled to get it in. The canvas, shrunken and stiff with salt, was tighter than a war drum. Finally we punched it into a small enough bundle to pass gaskets around it and lash the whole mess to the deck.

Wham! The bow leaped high into the air, then down it crashed. A solid block of water hurtled aboard. Clinging to the shrouds, our arms were almost wrenched from their sockets as the water dragged us down. Our fingers were bruised and bleeding. Coming aft, we furled the storm trysail. Even it could not have survived much longer. The deck was no safe place. Waiting for our chance, we shoved back the companionway hatch and dropped below into the after cabin. The last card had been played.

But this proved to be only the prologue.

Below, a state of chaos reigned. Wet clothes, books, and miscellaneous gear lay about on the floor and, as the ship rolled and pitched, the whole mess slid about in unison. Nothing was still. The brass lamp danced in its gimbaled bracket. The barometer swayed back and forth on its hook. A raincoat swung pendulumlike. A heavy lurch, and more objects jumped from the racks and joined the litter on the floor. To remain in a sitting position one had to wedge oneself behind the chart table or behind the companionway steps. It was hot, stuffy and damp. Normally undetectable, the whole cabin reeked with the stench of bilge. Water was everywhere. It sloshed around below the floor boards. It dripped down from the skylight. It churned about on the deck overhead. Frequently it struck staggering blows under the counter which made us flinch and look at each other anxiously. Even so, the noise was less than on deck and the shrieking of the wind sounded far off. One could at least talk below, even though every movement was an acrobatic stunt and to stay in a bunk one had to tie up into a knot. Given an opportunity to talk, there was nothing to say.

Bert's long form squeezed down through the hatch. His habitual smile was gone. "There's a helluva lot of water washing around down in the hold," he announced. "I wonder if that small leak by the sternpost has opened up any."

We cleared up some of the debris from the floor and raised the small square hatch at the foot of the steps. I looked down and suddenly felt sick. A goodsized stream of clear water was pouring in.

"Good God! The pumps! Quick!"

Even in the short ten minutes I had been below, a big change in the weather was apparent when I came on deck again. Squalls followed each other without a moment's break, and overhead was almost as dark as night. The wind had risen to a shriek and the lurching of the ship was so violent that it was all one could do to hold on. I crawled forward, gaining a partial shelter behind the weather bulwarks, and finally reached the mainmast. The air was charged with rain and spume; the wind beat back one's breath. Bracing myself with one hand clutching the main shrouds, I grabbed the pump handle with the other. Back and forth. Back and forth. Clear water flowed from the lip of the pipe. Only when there is a great deal of water in the bilges is it clear. Would the leak gain on us? What if the pump should jam?

The noise was terrific. Lines banged on the mast with the sound of riveting machines. The rigging hummed under the terrible strain like giant harp strings. The sea roared and thundered and above all was the deafening screech of the wind, raging with demoniac fury. Back in the fo'c's'le was shelter from this madness, but there was also the agony of inaction. On deck, unbearable though it was, one was at least actively, or even frantically, occupied. Below, there was nothing to do except watch that damnable barometer, speculate on the duration and fury of the storm, and wonder how hard the next wave would strike.

Sandy and Ed decided to have a look at the sea anchor and slack off the cable to counteract chafe. They fought their way forward, commenced to cast off some of the many turns about the anchor windlass, when suddenly they realized that there was no cable leading from the bow. It was a staggering, almost incredible, realization that that brand-new six-inch rope had parted in less than twenty minutes. It had not chafed through—the stub showed a clean snapping of the line at the chock. It was well-nigh impossible to imagine a force with all the give and resiliency of water breaking a line as strong as that. Without the drogue to hold the bow around, the ship lay in a critical position, beam on to the wind. At any moment a wave might break aboard, tear open the hatches, and fill her. We could not think of hoisting a riding sail, since, with the wind at hurricane force, no canvas could have survived. The leach of the mainsail was ripped from the gaskets and torn to ribbons. The sail was furled on the boom and the whole lashed down to the deck, but the end which projected beyond the gaff was caught and literally blown free from its stops.

We noticed then that the ship was not lying directly across the wind, as had been expected. Apparently its force was so great that the bare poles alone acted as sails, heaving her to with the wind a point or so before the beam. The rolling was now appalling. The lee rail dipped constantly under and the ship staggered with the weight of water on her deck. To run before the storm with that leak aft would be unthinkable.

Below, we tried to shift some of the movable gear in the hold to the windward side, in order to counteract the sharp list caused by the wind. Work in that stuffy hold was a nightmare. It was inky black, the light machine had given out, and of course there was no way of lighting the place. Four of us, stripped to the waist, struggled over tumbled gear with a half-dozen bags of coal. A bag of coal is a clumsy object to handle under the most felicitous circumstances; in this rocking, rolling, pitching hold, full of innumerable things to trip over and collide with, they were almost impossibly unwieldy. During this struggle a particularly violent lurch wrested a five-gallon can of coal tar from its lashings. The can burst against the mast and we were soon covered by the sticky, burning tar. But somehow under circumstances such as these, extraordinary feats can be accomplished, and eventually the coal bags and some other heavy gear were piled and wedged over on the starboard side. To further throw the weight on the windward side, we next decided to smash open the freshwater casks on deck. This would lower somewhat the ship's center of gravity, which might help a little. Bert and Dick seized axes and struggled up on deck. Six good blows and most of our precious drinking water was gone. No matter, perhaps we would soon be beyond the need of it!

Taking turns, we were pumping now every fifteen minutes. To stay on deck was a considerable feat and an experience which I shall never forget. From the after cabin the first move was to slide back the hatch—cautiously because green water was coming aboard all the time. You thrust your head through the opening and received a resounding buffet, then crawled up slowly over the closed doors, and, holding on to anything in reach with both hands, you flopped to the crazily heaving deck. The force of the wind was such that it was practically impossible to stand up. On hands and knees, never for a moment relaxing a firm hold on something, you worked your way forward behind the partial shelter of the weather rail. The main shrouds amidships broke the wind sufficiently so that here you could stand. Next, the pump handle had to be extricated and fitted into its slot; then, with one hand grasping the shrouds and the other on the pump, you set to work.

Meanwhile, what a spectacle was taking place before your eyes! Above, clouds tumbled over themselves in confusion as they hurried northward. Most of them were low and very dense, but occasionally they would open up and through a sort of tunnel of clouds would flash a patch of bilious yellow sky. At one time a comparatively large break revealed two or three high distant clouds. They, more than anything near at hand, indicated the incredible speed of the wind. Everyone has watched a distant airplane traveling at, let us say, seventy miles an hour, and wondered that it moved so slowly across the sky. But these high hurricane clouds! They sped past at easily twice the velocity of a seventy-mile-an-hour plane.

Closer by—in fact, right on top of one—the sea was performing such spectacular acrobatics as are never seen in any but a cyclonic storm. So terrific was the force of the wind every now and then that an area of huge raging waves, many acres in extent, would be suddenly flattened out as if by an enormous invisible paddle. Again and again huge cross seas collided head on, shot high into the air, and the tremendous mass of water was plucked bodily from the ocean and hurled several hundred feet by the wind.

On deck the law of gravity was blithely scoffed by every rope end— following the wind, they stood out horizontally, pointed upward or downward, directed not by the ordinary laws of the universe but following blindly their new master, the wind. On the starboard shrouds hung a ring life preserver to which was attached a flash can. Somehow this had broken its lashings and hung from the ring on a four-foot length of line. This can must have weighed three or four pounds, and yet it, too, stood horizontally out from the rigging. Such an occurrence does not seem astonishing at the time, but afterward, when you try to estimate the force of a blast which could accomplish such a feat, you are appalled.

The storm raged on with unabated fury. It was now impossible to face the wind. No demarcation existed between sky and sea. Wind and water were one, fused into a strange, fiendish element that tore at the *Chance*, intent upon her destruction. Blindly the little ship fought on. There was nothing we could do to help her. I marveled at her struggling beauty. I remember being vaguely surprised that she had lasted as long as this. Falling into a deep valley between the mountainous seas, the ship was almost in a calm, and then, as she rose on a breaking crest, the wind attacked her as if to tear her to pieces. Just under the lee side lay a patch of smooth water caused by the drift, the one tranquil spot in a universe gone mad. It fascinated me, and the temptation to slip in here and get it over was almost insurmountable. Anything to escape the noise. Did one have to be annihilated with such a wild accompaniment?

Fortunately all hands were below when the first knockdown came, else someone would have been washed overboard. The ship shuddered convulsively under the terrific impact of a monstrous wave. Then over she heeled. We were thrown from our positions as the floor sloped up at a crazy angle. Down she went, farther and still farther. I was squeezed in by the forecastle table. Looking out the skylight, I saw solid water rushing by. Still farther down she was pressed. The floor was upright, like a wall. A suitcase dropped plump from a bunk on the starboardside to the corresponding port one. On her beam ends the ship quivered. Would she come back?

Seconds passed like aeons. Events of my life flashed in review.

Then slowly she righted herself, staggering, fighting against the wind and water which held her in a viselike grip. I opened the hatch and scrambled out on deck. No use being caught like a rat in a trap. The first thing I noticed was that the big wooden sea chest, which had been fastened beside the engine-room hatch, had vanished. Its heavy iron handles, bent and twisted like bits of wire, remained lashed to the deck. Loaded with stops, mooring lines, and all sorts of spare gear, it was no light object, and yet a wave had picked it up as if it had been a chip. Had it but touched the rail in passing, it would have certainly stove it in, but there was not a mark, not even a scratch. The iron handles alone gave mute testimony to the power of that sea.

Unbelievable as it was, the wind continued to increase. Overhead it shrieked and screamed without a break. Waves, one after another, struck the trembling hull. The timbers creaked and groaned. Again and again the deck canted at crazy angles as floods of spray whipped by. Crawling blindly along the deck, I made my way back to the after cabin. And again came the agony of waiting. All one could do was sit, braced as securely as possible, and think about eternity with a poignant sensation of fear and excitement gnawing at one's stomach. There was little conversation. It was impossible, of course, to speak what was in everyone's mind, and any other remarks were palpably hollow. Bert and Dick were up in the fo'c's'le. Here in the after cabin there were the remaining five of us, the same five who had been through those storms off the Australian coast on the way down to Sydney. At the time we had thought them to be severe. Looking back, they seemed child's play by comparison. Skip sat with the barometer held in both his hands. He stared at it, fascinated, as a soothsayer stares at his crystal, and uttered over and over in awed tones, *"La Sorcière, qu'elle tombe, qu'elle tombe encore!"*

We were all nearly out of our minds, I suppose, feeling as prisoners feel awaiting the shooting squad. There was present a curious mixture of fear and bravery. We were afraid, desperately afraid—only an imbecile would not have been—and yet there was no question of breaking down. If there were anything to be done, we welcomed it, even though it meant a greater risk on deck. The barometer was now 29:25, the pointer swinging wildly as we tossed from crest to trough, so quick and violent were the changes in atmospheric pressure near the center of the storm. No one spoke. Stubbornly the *Chance* was hanging on. We almost wished the end to come swiftly.

Then suddenly the ship was again flung over on her beam ends and the topmasts were washed by the raging seas. . . . Slowly she came back, groaning under her terrific burden. I knew she could not take this punishment much longer, yet hope tortured me. She had come back twice! Before, I think I had been resigned. The barometer almost touched bottom. The pointer swayed so much that one could hardly read it. One second it was down to 28:85, the next up to 29:20. The noise was deafening—wind shrieking, spray and rain hissing, waves thundering against the hull. The motion was bruising and relentless. Half stunned, we waited and watched and listened and thought.

At four o'clock the glass started an upward trend. Wind and sea raged on, unabated, and visibility ended in a gray wall scarcely a hundred yards away from the ship. The storm was now striking from the north, and colossal cross swells pounded her from all directions. Three more hours, each one an eternity, passed as the squalls became fewer and with longer intervals between them. Occasionally clear spaces appeared through the clouds and stars flashed out, cold and distant, gleaming down on the heaving ocean. All hands took turns at the pumps. The little ship had ridden out the storm.

The watch was set. At midnight the glass had risen above the 30:00 mark. I struggled to my soaking bunk. All night the ship lay under bare poles while her crew slept the dreamless sleep of exhaustion. By morning the wind had fallen to a light northerly, but the sea was still enormous. Heavy swells rolled in from all directions and, although they were no longer dangerous, the tossing and pitching made it almost impossible to stand upright. The leak, although still a good-sized stream of water, had diminished now that the counter of the ship was not being subjected to severe blows. By pumping twenty minutes out of every hour we were able to keep it in check.

Below, the *Chance* was an appalling sight. The ladder leading down from the galley hatch had somehow come adrift and worked its way well forward into the fo'c's'le. A barrel of flour, more than half full, had snapped its lashings and smashed itself against the stove. Next, a wave had come plunging down the hatch and the whole galley was a mass of paste. In the absence of the ladder it was necessary to hang at arm's length from the hatch and drop to the galley deck. The drop was a matter of only six inches, but that was enough in a widely swinging ship, and not one of us came in through that hatch without sitting down abruptly and hurtling about for several minutes in a sea of paste. A touch of gaiety was added by a bottle of ketchup which had leaped from its shelf into the melee. Stove lids, broken crockery, cinders, and so forth contributed to the confusion. In the fo'c's'le bedding, personal objects, and wet clothes were scattered about. The hold was even worse. The paint and tar that had come adrift covered everything from spare sails to rubber boots. This mixture was tracked from one end of the ship to the other. We poured it out of our shoes and pumped it up from the bilges. It spread all over the deck and streaked the newly painted hull, so in the end the ship resembled a mud turtle crawling out of a swamp.

On deck we took stock of the losses. Among other things, the sea anchor, a patent oil spreader, a sea chest, complete with all gear inside, some spare lumber, a boom crutch, and the precious keg of Madeira wine, which we had contemplated smuggling into New Bedford, had all departed. Of the sails, the foresail was ripped but could be mended; the main topsail had been blown to pieces and the leach of the mainsail was torn beyond repair. The latter could, however, be carried with a reef set in it. Our six water barrels were hacked open and the contents lost, but fortunately the tanks in the hold still held fresh water, which, however, when we came to investigate, was chocolate color. All the spars appeared in good condition and I was thankful that we had not resorted to cutting away the masts during the height of the storm, a practice sometimes followed when the wind pressure aloft is so great that the boat is in danger of capsizing. Last but not least, both bilge pumps functioned. All in all, we had come out lightly.

From the direction of the wind and the condition of the sea, coupled with the rapid descent of the barometer, we estimated that the center of the hurricane had passed only three or four miles to the south of the *Chance*. There is a calm area in the center of a cyclonic storm and this region is the most feared by mariners, for the wind suddenly stops, leaving the vessel drifting helplessly, and then a few minutes later it bursts upon the ship with renewed violence from the other direction. I have often wished for a barograph record of our barometer's descent. It could not be estimated accurately, but we figured that in twenty hours it had dropped almost an inch and a half. With a normal daily range of four to five hundredths of an inch, this gives a fair idea of the violence of the storm.

The day was clear and fine and a warm sun was shining. Were it not for the enormous swell, it would have been very pleasant. Sail was set, and after a hot breakfast we began to feel almost human. All day we toiled, trying to get things ship-shape once more. Clothes and blankets were spread out to dry, gear straightened, and the quarters below made more livable.

The following day, although there was a slight easterly swell, the sea was almost normal. The breeze had come up from the south after a calm and the old ship trundled along well, despite her shortened canvas. She continued to leak, however, and we pumped out three to four hundred strokes an hour. That night a small squall brought more wind and we found ourselves anxiously regarding the barometer—"the Sorceress," Skip had called it—lest it give some sign of a repetition of the storm.

Our fears were unfounded. Dawn ushered in another clear, flawless day, with a gentle southwest wind on the beam. The ocean had now lost its pellucid blue and there was a chilly tang in the air. The Gulf Stream had been crossed. The home port lay just under the horizon's rim.

⚓

Some Sea Chanties

Whiskey! Johnny!

(Halyards)

O WHISKEY is the life of man,
Whiskey! Johnny!
O whiskey is the life of man,
Whiskey for my Johnny.

I drink it out of an old tin can,
Whiskey! Johnny!
I drink it out of an old tin can,
Whiskey for my Johnny.

I drink whiskey when I can,
Whiskey! Johnny!
I drink whiskey when I can,
Whiskey for my Johnny.

I drink it hot, I drink it cold,
Whiskey! Johnny!
I drink it hot, I drink it cold,
Whiskey for my Johnny.

I drink it new, I drink it old,
Whiskey! Johnny!
I drink it new, I drink it old,
Whiskey for my Johnny.

Whiskey killed my poor old dad,
Whiskey! Johnny!
Whiskey killed my poor old dad,
Whiskey for my Johnny.

Whiskey makes me pawn my clothes,
Whiskey! Johnny!
Whiskey makes me pawn my clothes,
Whiskey for my Johnny.

Whiskey makes me scratch my toes (gout?),
Whiskey! Johnny!
Whiskey makes me scratch my toes,
Whiskey for my Johnny.

O fisherman, have you just come from sea?
Whiskey! Johnny!
O fisherman, have you just come from sea?
Whiskey for my Johnny.

O yes, sir, I have just come from sea,
Whiskey! Johnny!
O yes, sir, I have just come from sea,
Whiskey for my Johnny.

Then have you any crab-fish that you can sell to me?
Whiskey! Johnny!
Then have you any crab-fish that you can sell to me?
Whiskey for my Johnny.

O yes, sir, I have crab-fish one, two, three,
Whiskey! Johnny!
O yes, sir, I have crab-fish one, two, three,
Whiskey for my Johnny.

Leave Her, Johnny

(For Pumping and Halyards)

I thought I heard the captain say,
Leave her, Johnny, leave her;
You may go ashore and touch your pay,
It's time for us to leave her.

You may make her fast, and pack your gear,
Leave her, Johnny, leave her;
And leave her moored to the West Street Pier,
It's time for us to leave her.

The winds were foul, the work was hard,
Leave her, Johnny, leave her;
From Liverpool Docks to Brooklyn Yard,
It's time for us to leave her.

She would neither steer, nor stay, nor wear,
Leave her, Johnny, leave her;
She shipped it green and she made us swear,
It's time for us to leave her.

She would neither wear, nor steer, nor stay,
Leave her, Johnny, leave her;
Her running rigging carried away,
It's time for us to leave her.

The winds were foul, the trip was long,
Leave her, Johnny, leave her;
Before we go we'll sing a song,
It's time for us to leave her.

We'll sing, Oh, may we never be,
Leave her, Johnny, leave her;
On a hungry ship the like of she,
It's time for us to leave her.

Rolling Home

Fare ye well Australian daughters
Fare ye well my blue-eyed maid
For today we have our orders and
For home our course is laid.
Lay aloft and loose your topsails
Overhaul your running gear
Lend a hand and man the capstan
See the cable runs down clear.

Rolling home, rolling home,
Rolling home across the sea;
Rolling home to dear old England,
Rolling home, dear land, to thee.

Eastward, eastward, ever eastward
To the rising of the sun;
Homeward, homeward, ever homeward,
To the land where I was born;
Where the dear and patient mother
Sitting waiting there for me
With a welcome home to England,
Welcome home across the sea.

Chorus

Up aloft amid the rigging,
Sings the loud exultant gale,
Like a bird with spreading pinions
Flies our ship before the gale;
While the wild waves cleft behind us,
Seem to murmur as they flow,
There are loving hearts to greet you
In the land where you would go.

Chorus

O, a thousand leagues behind us,
And a thousand leagues before,
Ancient ocean heaves to bear us
To that well remembered shore.
So we'll raise our happy voices
In the watches of the night,
And we'll see the land of England
In the dawning of the light.

Chorus

A Runaway Chorus

What shall we do with the drunken sailor?
What shall we do with the drunken sailor?
What shall we do with the drunken sailor?
    Early in the morning.

    Way, hay, there she rises,
    Way, hay, there she rises,
    Way, hay, there she rises,
    Early in the morning.

Chuck him in the long-boat till he gets sober,
Chuck him in the long-boat till he gets sober,
Chuck him in the long-boat till he gets sober,
    Early in the morning.

    Way, hay, there she rises,
    Way, hay, there she rises,
    Way, hay, there she rises,
    Early in the morning.

Lock him in the guardroom till he gets sober,
Lock him in the guardroom till he gets sober,
Lock him in the guardroom till he gets sober,
    Early in the morning.

Way, hay, there she rises,
Way, hay, there she rises,
Way, hay, there she rises,
Early in the morning.

Blow the Man Down

Oh blow the man down, bullies, blow the man down.
To me Way-ay, blow the man down.

Oh blow the man down, bullies, blow him away.
Oh gimme some time to blow the man down.

We went over the Bar on the thirteenth of May.
The Galloper jumped, and the gale came away.

Oh the rags they was gone, and the chains they was jammed,
And the skipper sez he, "Let the weather be (hanged)."

As I was a-walking down Winchester Street,
A saucy young damsel I happened to meet.

I sez to her, "Polly, and how d'you do?"
Sez she, "None the better for seein' of you."

Oh, it's sailors is tinkers, and tailors is men.
And we're all of us coming to see you again.

So we'll blow the man up, and we'll blow the man down.
And we'll blow him away into Liverpool Town.

⚓

35

Rendezous

CHARLES RAWLINGS

AT FIRST we said nothing about Pete Howard. Then, after a few weeks, "Any word from Pete Howard?" we wondered. Finally one day I sat across the hexagonal oak table where we always sat in the Yacht Squadron wardroom and sucked a tooth at big Charlie Nichols. "Pete Howard's dead, ain't he?" I asked.

The big harbor was sparkling in a crisp, cold February westerly. It is an important convoy port. "An Eastern Canadian port," will let you by the censors with it. Its wide mouth opens east to sea. Across its narrow neck stretches the submarine boom. Its upper reach is a broad deep anchorage where the rusty anchor chains of fifty deep laden freighters rumble sometimes in a single day's mooring. The Yacht Squadron where we sat was on a small bluff that overlooked the harbor narrows and eastward out to sea. Out there our courses used to be. We'd sailed them hard in the peaceful years that seemed so long ago, when they were clean bold water instead of mine fields. Pete had sailed them better than anybody. He was a splendid character, I had always thought. I would use him sometime, I thought, in a story.

There was something about him like Lucius Quinctius Cincinnatus, the Roman. He was peaceful and contented sailing one of our squadron's little Maritime one-designs. They were eighteen-foot jib and mainsail sloops William J. Roué had designed for us. Pete's was named *Sue*. He could have afforded anything he wanted and sailed it to win. We had a fine fleet of Eight-Metres and a big cutter class and he could have dominated either one of them but he was content with the little *Sue*. She even slept with him. He'd sail her up the harbor in the early evening after the racing and drinking was over and hang her to her home mooring off the big Victorian, red-brick Howard mansion on the upper basin. When we needed Pete for the big racing—to be captain of the Eight-Metre team that met a team from the Eastern from Boston every two years or skipper a defender for an important cup series, he'd toss the *Sue's* sheets up over her boom and put a hank of gasket around her jib the way Cincinnatus tossed the lines over his plow handles, and come along to do battle for us.

Charlie Nichols shuddered, shaking the red tabs on his tunic collar. Charlie is a Colonel of Engineers.

"Who knows if Pete's dead?" he said. " 'Two of our aircraft did not return!' That's all anybody has heard so far. 'Two of our aircraft did not,' said the Air Ministry, 'return.' Pete was at the tiller of one of them. Down in the Ruhr? Down somewhere? That does not mean he's dead. Who can tell? Who—— knows——?"

His voice dwindled off and his eyes stared out over my shoulder up the harbor and his mouth slowly gaped under the Anthony Eden moustache he wears. The starkness of his eyes made me turn too and there was the *Sue*. A Norwegian freighter was moored directly in her course and it seemed from our angle that she was going to run her down. Two small pea-jacketed figures on the freighter's after deck started running and we held our breath. But she disappeared behind the big rusty hull and after a moment her tiny bow stuck out and she came on. The breeze was a bowling westerly as I said and it was right on her tail. Tide had her strongly too. Halfway up the stay her jib filled and backed and then jibed and filled again. It had worked itself loose from the gasket and flopped its way up the stay itself. You could see a loop of loose halliard swinging. Over her bow she trailed the broken end of her mooring line like a run-away puppy who had broken his leash. Her canvas was sooted and gray and there was a pitiful small windrow of snow on her tiny bow deck. Steadily, determinedly she came on bound out—bound for sea.

"Say," breathed Charlie, "that poor little crate. Didn't anybody think of hauling her——. Damn it, she's goin' some place, all by herself. Look at her, comin' along straight and true. That's the proper course baby," he said to her, "that's where Pete bears all right——."

Then he stopped and I could hear him breathing as we followed her jogging along so steadily, so determinedly. I rubbed the frost off the easterly window and we squinted at the gate boats on the submarine

boom. The gate was open. There was something ready to go out probably, far up the harbor and the gates were open for that. The *Sue* was
clear for sea. Then Charlie got up and I could hear his boots slowly
creaking as he walked across the room and out the hall to the telephone.

"Yes," he said, "this is Colonel Nichols speaking. Get me through
quickly, please. That you, Major? Nichols! Here's a change of orders
on that eastern mine field test. We were running it tomorrow, at nine.
We'll run it today—right now. Look out your window. There's a small
sloop running out toward you. No, there's no one aboard her. She's
broken her mooring and sailing herself. That southerly set of tide will
bring her down just enough. Let her go. *No I don't want to save her.*
Let her go, I said. Let her go merciful and swift, pray God. How's she
comin'———?" The rumbling deep throated "bo-o-o-oom" of the mine
came then, almost a musical detonation cleansed by a half mile up
wind. "Yes," said Charlie, "I knew her. She belonged to a friend of
mine. No, he won't be sore. Quite otherwise, I think. He's dead."

⚓

36

Little Ships Are Safest

ALAN VILLIERS

OFF the North Cape of New Zealand I saw a small ketch about 31 feet long, wallowing there in a gentle breeze. She was the American yacht *Igdrasil,* bound from Auckland to Invercargill in the south of New Zealand, and she was being sailed by a Maine school teacher and his wife around the world. She was heading the wrong way to reach Invercargill from Auckland, to be sure, for she was heading north; but she very likely made it.

At Auckland, when I came in, the first thing I saw was another ketch, 29 feet long, three years out of Seattle on a voyage around the world. She had been dismasted in a hurricane near Samoa and was in for repairs. Her youthful owner, whose only comrade was a bronzed Tahitian mariner, announced casually that he was going on to Sydney and the islands, and Singapore, and so on to the westward home. It might take six years.

The multitude of these wanderers who, until the outbreak of the war, were sailing in the South Seas was extraordinary. At Norfolk Island, lonely outpost tilled largely by transhipped descendants of the *Bounty* mutineers, I saw a Norwegian cutter on the reef. Her name was *Ho-Ho,* to express the merriment of her carefree mariners at the idea that she might sail around the world—for when their plans were first announced the roars of the waterfront populace of Drammen in Norway were ribald and loud. "Ho ho!" they shouted with derisive joy—and so the three mariners called their fifteen-ton ship *Ho-Ho* and sailed out of Rio de Janeiro and Buenos Aires and through the West Winds from there to Hobart in Tasmania, and from there by devious routes among the islands, touching on Norfolk's reef; and onwards again to Callao in Peru.

At Tulagi in the Solomons I saw a converted lifeboat that had been sailed there from Staten Island, by a single-handed wanderer over three score and ten. True, he unfortunately came in dead; for he had died at sea a few days before coming to the Solomons, and his little vessel had drifted on a reef off Ysabel with her American flag at half-mast, where the old man had hoisted it before he went below to die. There was little or nothing to eat on board, and the vessel was ill-equipped with navigational aids.

Tragedy sometimes is shipmate on these strange voyages. There was the German yacht which put in at Barbados the other day a month out from Las Palmas, and though three had sailed with her only one arrived. A German student, he explained that he turned out one morning to take his watch, and his two companions were not there. He sailed back many miles along the way he had come, and saw nothing. So he came in alone.

And the mystery of the disappearance of that most strange of all sea-wanderers, Ira C. Sparks, of Indiana and Honolulu, who sailed a sixteen-foot boat out from the Hawaiian Islands towards Mecca one day in 1924, and never was seen or heard from again, though his frail punt was found seaworthy and whole at Zamboanga in the Philippines months afterward, has never been solved. His craft was the *Dauntless,* and there must have been much that was dauntless about the mysterious farmhand navigator, who though he did not come in with his ship, made a small-boat voyage that even in this day is extraordinary.

Whence come all these strange wanderers, and why? They are not sailors. The professional seaman rarely turns to the sea for escape. Few of them are yachtsmen; many of them, before setting out on their always long and often hazardous voyages, have had little or no experience of the sea. Yet they get by—mostly. Sometimes they do not. The number setting out with great plans is greatly in excess of the number returning with something accomplished. Almost every day of the week, some small craft sets out on a "world" voyage from somewhere, though often the sea stores consist mainly of a fishing-line with the use of which the owner-master is not too well acquainted, and the navigating instruments are an uncompensated compass and a run-down watch.

There is a more or less beaten track for the ordinary circum-

navigation—down from New York or Gloucester or Norfolk or somewhere, and through the Caribbean, with the hope not to find any islands
unexpectedly grating beneath the keel. Then through Panama and on
to the Galápagos and before the southeast Trade to the Marquesa and
the Tuamotus—though the wise pass these up. A mighty bad place for
any kind of ship, round there. Tahiti, then and westwards through the
South Seas with leisurely calls at as many islands as possible.

And so to the more primitive Melanesian islands, the New Hebrides
and the Solomons, and the wild outposts of Northern New Guinea;
and onwards through the Dutch East Indies and westwards before the
Trade of the Indian Ocean, and westwards still around Good Hope with
the favoring Agulhas current (though there is apt to be wild weather
here) and so home with the Trades of both Atlantics. It is an easy run,
once Hatteras is passed. The bad seasons can easily be avoided, for the
simple reason that their habits are well known and are listed in books.
In this manner even a very small vessel may safely be sailed around
the world. Many have been, and the defeat of the ocean by small, well-
found yachts has become commonplace.

Which brings me to the costs of this kind of thing. One does not
need a full-rigged ship. A bald-headed diesel schooner or ketch will do
just as well, something about 40 feet on the water line, or maybe 50
or 60, if you want size, husky and staunch and seaworthy and strong.
You can—or you could—buy such a vessel, fit for long voyaging, for a
very few thousand dollars, if you knew where to go and did not pay
undue attention either to the optimism of brokers, or their pessimism.

You ought to be able to do without both gasoline and kerosene, and
for that reason diesel power is preferable. It is a big reason. You ought
to be able to do without a professional crew. If you cannot handle your
craft yourself, you ought not to go. If you cannot get a few friends to
come—the fewer the better, and for God's sake watch the women, if
they *must* be along—better go alone. If you must have a crew, accept
nobody's recommendation of anybody. It doesn't count. You will have
to have someone you can get along with *yourself*. I have a preference
for Nordics, myself, or for Rapa Islanders and pure-blooded Tahitians
and boys from the Tuamotus. There is scarcely an adult white anywhere who hasn't some fatal disadvantages that show up in the course
of a long cruise in a small vessel.

You've some yourself. That's all very well, if you know it. But for heaven's sake, choose your shipmates carefully. One bad one can easily spoil what might otherwise have been a splendid voyage; and you have to bear in mind always that a group of human beings collectively seeking escape at sea is apt to throw up the morose and the discontent when the inevitable discovery is made—and it is made quickly—that there is no "escape" merely because there is water all around you, and not land. You take your problems where your carcass is, and if some are left behind others will take their places, maybe worse ones.

But you probably will not have to bother. One of the most difficult things about your voyage—if you really are bent on making one—may be to get any shipmates at all. I owned and sailed the *Joseph Conrad* for close on three years and in the course of that time covered some 60,000 of the most interesting sea-miles I knew; yet I carried no friend with me that voyage. A few came, for odd stretches; none stayed. You can't expect people—grown-up people, with responsibilities and jobs, and homes and things—to change their lives merely because you buy a ship.

Besides, there's a lot to put up with at sea. If the weather is always good, it palls; if it's bad, lots of people get bad-tempered. The continued motion of a small vessel is hard to put up with. One atoll, after all, is very much like another; and there are plenty of coconuts in Florida. And sometimes inexperienced mariners take badly to any notion of disciplined command.

As for navigation, there was a time when I'd have said this was important, when I might have intimated without hesitation that at least the elementary knowledge of the simpler methods for ascertaining latitude and longitude was highly desirable. I still think it is, but I don't know. This last voyage I met some fellows taking a craft around without even the proper nautical tables on board, without accurate instruments, without proper charts, and with only the vaguest idea even of how to tell whether their compasses were in error. Yet they were getting there. God is kind. I saw two of them in Tahiti, which after all is a mighty high island, and on a clear day you can see the place a hundred miles at sea. But it would be awkward to sail by in the rain.

But a sufficient knowledge of navigation may readily be acquired. There is nothing very difficult about it, especially with the latest formula and simplified tables which the benevolent Hydrographic Office of this country supplies. You can learn enough in two weeks, if you've a reasonably mathematical brain; the rest you'll pick up—or else. The Coast Guard is a mighty fine institution. I think maybe the reason why one sees more than twice as many American small yachts sailing the seven seas as those of any other nation, is largely due to those two noble government departments, the Coast Guard and the Hydrographic Office.

No other country looks after seagoing amateurs, tyros, and nitwits as does this country. If you get in a jam and you're not too far away— and you can go a mighty long way—you've only got to holler for the Coast Guard; and if you've gone in some place where you'd no earthly right, and you've neglected to visit a port of entry or given an account of your vessel, her stores, her crew and all that—you're doing it for "science." But you'd be well advised to have those alleged scientific objectives well documented.

Seamanship? It takes no superman to get a small craft around the world. That has been well proved. A staunch small vessel can safely be taken anywhere, and back again. There are regular transatlantic races for small yachts, races from the West Coast to Hawaii, races across the stormy Tasman Sea, to Havana, to Bermuda. There are seldom any accidents. If you get into a jam, a hundred to one it's your own fault. And it's a thousand to one, or thereabouts, that you'll get out of it safely.

Some of the craft I have seen ambling about the world have been crazy things that I would not have taken cheerfully for a summer afternoon's sail on Long Island Sound. Some of the lone "navigators" I have met had a mighty queer look in the eye. But they get around, and that's something. They sail, and they come in again, and that is all they aim to do. Some of them still made the most elementary errors of seamanship, but they get by. God is kind, and the way of a staunch little ship in the great sea is wondrous to behold. They've only to keep off the land. . . .

I remember seeing a piece in the press—"Five to Circle World in Schooner"—and in the same paper another headline mentioned "Ocean Voyage in Canoe." Both of these were probably more plans than happenings; but they became voyages, like as not. But the strangest mariner of all was surely that apprentice in the Swedish schoolship who stole the captain's bathtub and sailed himself ashore in it somewhere in the Torres Strait Islands. Then the government picked him up and deported him because he did not have the proper visas, and he did not bring his craft to a port of entry!

It can be a hard world, even for the intrepid.

⚓

37

Old Man Noah Had a Boat

W. E. PHILLIPS

It RAINED that day, not just a gentle rain, but a steady downpour. I drove down to the yacht club with some fittings I had repolished and from the locker room I wandered into the lounge. Seated by the fireplace was our old retired Professor of Assyrian Antiquities, smoking his pipe and listening to the radio. Some commentator was giving what he called "The story behind the news." When he finished we shut the radio off. "Funny thing," I said, "how those chaps can dig around and get the information that gives you quite a different slant on what you read." "I don't know," said the Professor, "I have done a lot of research in old manuscripts and it would surprise you what you can find, if you are lucky. Why I found out lots of things; things I couldn't even tell my class; they wouldn't have believed me. For instance, this rain today reminds me of Noah and the Flood; there's a story that follows that, that few people know or have read. You remember how the story of Noah and the Ark stops short when the flood went down and the begats begin, you know 'Shem begat Aram and Aram begat Uz' and so on. I was fortunate enough one time when I was in Syria to get an opportunity to meet an old monk, one of several who lived in a monastery near, I just forget the name of the small village, but it was not far from Mount Ararat. After supper one night they brought out some old, old parchments and I started in to read bits here and there when suddenly I ran across something which was of absorbing interest to me, as a yachtsman, who had done considerable cruising and racing and as an antiquarian.

I found that when Noah went on that famous cruise, with his wife and his three sons, Shem, Ham and Japheth one way and another they

Reprinted by permission of *Boating Magazine*.

were at it all season. It was Noah's idea, just a nice family cruise. Shem, Ham and Japheth had only been married a short time, a matter of some 40 or 50 years you may recall and had no family, so it was just like a honeymoon to them. Noah wasn't really old, not as they figured in those days, a mere matter of 600 years, I believe, but he liked to have the boys along to help handle the boat and of course their wives helped Mrs. Noah in the galley, so it really wasn't much work for any of them.

Noah had taken his time about building the boat, about 100 years or so. Having been working on it for some time he wanted to take a summer off and have a good long cruise. The flood came along while they were loading the provisions but it saved quite a lot of work as they did not have to launch the boat but just waited till the water rose and floated the boat off the stocks.

She was well designed, what we would call a 6 beam boat, 300 cubits long with a beam of 50 cubits. She had lots of draft, when loaded, and was built of gopher wood, properly caulked with linen tow and all seams tarred inside and out. After the first few days she took up and they seldom had to pump her.

They started on this cruise and she was well stocked so they had plenty to eat. Fresh water was no trouble, as it rained every day and all they had to do was catch the drip from the eaves. She had no great speed as Noah had just put on a cruising square rig. He had some light canvas but the first part of the cruise he never had a chance to bend it, as it rained every day. No one worried much; Mrs. Noah was either in the galley or doing a bit of knitting or mending; Noah puttered around the deck and kept things ship-shape, entered up the log and so on. As I mentioned before, the rest of them were on their honeymoon.

After six weeks of wet weather it cleared and Noah got up his light canvas and started in seriously to sail. After two months of cruising it was getting near the end of the season and they were on the Gilion River so Noah ran on down to the mouth of the river on the Caspian Sea and laid her up for the winter.

By next season the boys and their wives could not go cruising as little Gomer and Cush and Elam had come along and it was a busy season in the vineyards. Noah had planted some exceptionally good varieties on the mountain slopes and they were coming well into bearing. He was a connoisseur of good wine, and put down a few casks for his special use of Year 602 vintage, but Mrs. Noah would not let him keep more than a cask or two around the house, so he took the rest down and kept them on the *Ark*. He set a precedent since followed by many yachtsmen.

They all had large families and by the time the youngest was about 50, Shem, Ham and Japheth had quite a bit of time on their hands. The *Ark* was getting a bit old too but Noah kept her in good shape and every 25 years or so he put in a new plank or retarred the bottom. Gomer, Cush and Elam used to like to go down with Grandpa Noah and play around the boat, but they never got into the spirit room; Grandpa kept the key to that.

Shem, Ham and Japheth decided to each build themselves a boat and determined to build their boats of cedar. They argued long with Grandpa Noah about the subject but he had no use for cedar. He pointed out how with gopher wood he had only replaced ten planks in 100 years, and the old *Ark* was still sound and tight. How could this soft, brashy cedar compare with good sound reliable gopher wood.

Well, you know how boys are; wouldn't listen to the old man, and he was getting on in years, about 900 then, and they went ahead and built their boats of cedar. Sent away over to Lebanon to get it, too. They didn't build them as long as the *Ark,* so having large families, they increased the beam, to get the cabin space they needed.

While they were building, a traveller from the East came along and interested them in a mixture of rare gums and distilled wine. By this time the vineyards were producing well, so well that they had been obliged to find a new use for the wine. Shem had done some experimenting and had developed a distillation process. This distilled wine was potent stuff, but they found a ready market for it and made money out of the by-product.

They tried some of the traveller's mixture of gums and distilled wine and put on with a camel hair brush it certainly produced a fine glossy finish. But Noah would have none of it. He had used tar for 250 or 300 years and it was good enough for him. How much good would their fancy stuff they called var-nish be in salt water? All right, maybe, in a river like the Gilion, but what about sea-water?

In about 50 years they had the boats finished and launched, ready for use. Noah watched them sail but he refused to go out with them— not he—why, go out in one of those frail cedar things with only var-nish on the bottom, no sir, too dangerous; just foolhardy, that's what it was.

Most of his great-grandsons had gone in for the new sport of Chariot racing but Cush and Gomer were interested in model boats. They built them like the old *Ark* and square-rigged them the same way. But Cush was a careless, impetuous boy and one day when the rigging on his model came adrift he just tied it up again with the yard all askew. Grandpa Noah was sitting on a log watching the boys sail their boats and he reproached young Cush for his carelessness, but he was surprised to see Cush's boat start to sail to windward, while Gomer's boat drifted to leeward, as usual.

While Shem, Ham and Japheth praised the performance of their new boats, Noah was unimpressed. His heart was in the *Ark* and it quite literally kept his spirits up. He met all the raillery of his sons with the assertion that the *Ark* was still the better boat. One thing led to another until one day they challenged him to a race and he was forced to accept.

Having accepted, he retired to the cabin on the *Ark* to figure out his chances. True, he had a six beam boat against their four, but was a tar bottom really better for racing than a var-nish bottom? Then he had been using the *Ark* for many years as a special wine cellar, and he had more on board than he properly needed for ballast. He pictured the course, the likely winds and all the factors and he had soberly to admit that his chances were not so good.

Thinking of his ballast brought to his mind the cask of Year 602 he had been saving for so many years, so he went below and broached the cask and drew off a large dipperful. It was good, exceptionally good, and it stimulated his mental processes. One question became clear in his mind, if he could only in some way get to windward better than his sons, the race was his. Then his mind travelled back to little Cush's model with the cock-eyed rig. That was something, something worth a try, so, draining the dipper, he went ashore to consult Mrs. Noah privately. Several days' work produced a new sail and a new yard of extreme length was cut and trimmed; he was ready for a trial.

But Shem, Ham and Japheth were always around tuning up their boats, so he never dared hoist his new sail which he called a la-teen. The day of the great race arrived and on the third blast of the ram's horn, away they went. The first leg was to leeward, and as he feared, he lost distance rapidly. His sons cheered him derisively as they passed. But at the turn he dropped his squaresail and hoisted his la-teen and tried going to windward. She went; rapidly he overhauled them, passed them, and when the race was finished he was hours ahead.

So Noah died at the ripe old age of 950, a happy man who had proved that a boat may be old, even 350 years old, but she is still a good boat and a yachtsman is 'never too old to learn.'

⚓

38

I Build My Ship

DOUGLAS R. RADFORD

Not literally of course, for when only by intense concentration can I even saw a board in two along a ruled line, the possibility of my building a boat, even the smallest of boats, is as remote as—well, how remote are the stars, the really distant stars?

I stand simply amazed when some quite ordinary-looking person waves toward a nice-looking boat waiting to be launched and says: "Sure; I built her on the lot next door. . . . Oh, just my young brother and a friend of his," he replies to further question. "Made a steam box and bent all the ribs; set up a band saw in the garage. Took me nearly three years though."

Well, I suppose it's all true, but it seems impossible. Although I well remember watching Harry Pidgeon build his *Islander*. Month after month and year after year he hammered and sawed and planed—he even hammered himself an anchor and planed two masts—but then Pidgeon is quite an unusual person, one of the Seven Sages, for he has overcome what Man is born subject to; he has overcome Time. For Harry Pidgeon Time does not exist.

Of course, sometimes the boat that is waved at looks awful—and *is* awful. But imagine being able to build even an awful boat on the lot next door with just a few relatives and friends to lend occasional assistance!

There were plenty of times when I thought that a well-equipped boat yard with a hundred admittedly first class builders were going to fall down on the job—at least as far as I was concerned, unless I lived to an age far beyond the prescribed three score and ten. I visualized

"Bones," (the villain in this story), shrunken and bent and crabby, peering up from a wheel chair (the roll-yourself kind) at a creaking Willard feebly tapping his caulking mallet on the deck above and studiously separating his long white beard from the caulking cotton before driving it into the seams; and myself a mere shell of the splendid specimen I once was, squinting through double-thick lenses at neatly typewritten sheets of paper all headed "Job 30. Men's time and material used week ending . . ." and large sums of money shown at the bottom of each sheet.

Lots of fellows have said to me: "Well, I'm going to build a boat some day," or perhaps, less positively, "I'd certainly like to build a boat." But I notice most of them haven't gotten around to it yet. With a genial desire to help others get started, who wish to get started, I gladly relate for the benefit of all prospective boat owners how I built mine; or rather, how the building was brought about, continued and f-i-n-a-l-l-y finished—almost.

Of course, I've got to be careful; I've heard of suits for "defamation of character," libel suits and so forth, and as I expect to defame several characters before my ship gets herself built, a reasonable amount of restraint must be exercised.

Boats appeal to me irresistibly. I prefer them on the sea. But if I can't have them on the sea, I'll take them on a river, because all rivers lead to the sea. However, my first sailing was on the ponds of Blackheath, a cutter 17 inches long (I have her yet), built by my father, deadeyes, tiny wooden blocks, mahogany rail, a beautiful piece of work that I was too young to appreciate. But I sailed her consistently every Saturday. Prior to this boat a huge model three-master came to the house. As I remember her, she must have been a beauty. I could sit on the deck between the masts, and she was wheeled up to the Blackheath Ponds in a hand cart for sailing. My father and the man next door did all the sailing. I was permitted to rush around the pond, hoping each time to reach the other side before the boat did and watch ecstatically the "turning around and heading for the opposite shore" process. I soon gathered that this boat was for the grown-ups to play with. I was the excuse.

The big brown sails on the Thames barges sliding down past Greenwich held me spell-bound, and the occasional deep-sea craft towing up to the docks filled me with longing. Every foot of riverfront at Greenwich is rich with memories of famous ships and famous sailors. Even an ancient little pub looking out over the river is fashioned like the stern of an old fighting ship—if it's still there.

This is all the sowing of the seed that later germinates, grows and finally results in the building of a ship.

I have owned and enjoyed several boats, but, there was never a boat built that the new owner felt he could not improve upon. And so I drew a picture of the sort of boat I wanted. As a matter of fact, I was putting the finishing touches to the picture about 25 years ago, and then every so often I'd have to draw a new picture because the old one had gotten improved so many times it was hard to make out whether it represented a boat or "Salmon Ascending a Stream."

My drawings recorded faithfully the small but fairly regular increases in my weekly salary. Very early drawings showed an open boat with a dipping lug sail; the day I leaped from a weekly wage of $1.25 to $2.50 produced a picture of a sloop with a small cabin, plumb bow and a lot of outside ballast. There came also long periods when there were no pictures drawn at all, for shovelling river gravel at $2.50 per day is not conducive to yacht designing. There is a lot of fun, however, in drawing pictures of the boat one expects to have some day. It's a harmless, inexpensive pastime, and I recommend it unreservedly to all those who, having no boat wish to have one, or having a boat want a different one.

The condition of some of my earlier boats was such that I found it wise to establish pretty close contacts with the boat yards. Often on a Sunday night I would just manage to reach the sanctuary of a boat yard and leave my boat like a foundling on a doorstep for the proprietor of the yard to find early next morning, not omitting a note pinned to the mast saying, "Please take care of her."

Joe Fellows' was a regular Orphans' Home to my early boats, so when in trouble I just naturally headed for Joe Fellows. He was certainly a good friend to many a young boat owner in those days.

Struggling up the San Pedro Channel one day with a "new" old boat, I found to my dismay that I was not going to reach my usual haven. Choosing a likely beach to run in on, I espied a tiny new boat yard right abeam and for that I headed.

Well, I found a chap there they called "Bones." So I said: "Mr. Bones, will you haul my boat up alongside and repair the hole in her stern?"

Mr. Bones said he would, and looking down at the boat, asked, "Is that all you want done to it?"

That should have been sufficient warning to me. While I did not like his implication that a lot of other things needed doing beside repairing the hole in her stern, I did not realize I had fallen into the hands of a direct descendant of those creatures who used to walk a cow along the shore on a stormy night with a lighted lantern tied to her tail—and rifle the wreck in the morning. "Wreckers," they called 'em.

Leaving the exhausted boat leaning against some piling, I went thankfully home, and next morning Bones et al. mended the damaged stern and also pumped the boat out, shovelled out all the sand and mud with which she was filled, dried out the bedding, cleaned, scraped and repainted the interior because, you see, I had, in some peculiar way been psychically induced to leave my boat in such a position that at high tide the hole in the stern would be well below water before there was any danger of the hull floating.

When I found what had happened and received my bill, I immediately returned to Los Angeles, and driven by dire necessity, found other employment that would enable me to pay Mr. Bones more quickly. I have found it necessary to change employers a number of times since then, urged always by the same dire necessity.

Mr. Bones' yard was so strategically located and I fell victim so many times that soon I was toiling all the week to pay him his blood money on Saturday afternoons. It should possibly console me to some extent to realize that, built upon my toil and privation, the boat yard has become quite imposing in size and employs enough men to have strikes. Even Mr. Bones has changed in appearance. (The reader will appreciate a strong note of restraint running through this narrative—particularly at this point.)

After having lived in and around the boat yard for a number of years and gotten more or less inured to financing an enterprise of this nature, my boat had been in the hands of the yard so much that little could be found for the men to do.

How it came to pass I do not understand—I think I must be very easily led—but one day I found myself spreading out some of my more recent boat pictures in the office of the boat yard and asking questions.

A week or two later I remarked, in the same office and quite casually,

"One of these days I'm going to have a boat built just about along the lines of these sketches." That sounds casual enough, doesn't it?

I nearly collapsed when I learned that the boat was already started. "Go up in the loft; you'll find her lines are already laid down on the floor," I was told. I staggered up the stairs and sure enough there she was, all laid down. So that was that. Just like trying to make up one's mind to dive in off the very high springboard and then getting pushed in.

We discussed cost and methods of payment and such trivial matters, to a tune that was running through my head from Gilbert & Sullivan's "Iolanthe:" "Faint heart never won fair lady, never won fair lady," and I certainly intended winning the fair lady that was beginning to take beautiful form in the workshop upstairs. "In for a penny, in for a pound," the song goes on. Well, I was apparently following the theme of Mr. Gilbert's song pretty faithfully.

My advice to the man who intends to have a boat built is to arrange for a six-months' leave of absence, unless of course he is already a gentleman of leisure, and while he may prefer to return home to sleep he can at least have his lunches at the yard and not waste valuable time running about for meals. It is intensely interesting to watch any boat grow into form, to see the keelson bolted to the metal keel, the frames springing up and out from the keelson, the carlins tying them together, the frames shaping the skeleton boat and giving promise of her future grace and beauty. But when one realizes that this growing fabrication is a dream of years materializing right before his eyes, recognizes the lines as they fair up and the form as it takes shape, then I guarantee one's pulse will quicken and the boat yard will become a most desirable place in which to be.

To enjoy the building of your boat to the utmost you should be not quite able to afford her. The feeling that you really should not be engaged in such a wild adventure, and yet are now well embarked, adds zest to living. It also stimulates the builder to his best efforts; he too knows darn well you can't afford it, and the thought that he may eventually have to keep her himself is undoubtedly lurking at the back of his mind—or should be!

After the lines were truly laid down, followed the casting of the keel —quite a business. Then I watched them start whittling on a magnificent piece of camphor wood for the keelson. It alarmed me to see how that piece of timber became smaller and smaller, but at last it was reduced sufficiently and bolted down on to the keel. That was a most exciting operation. Then the stem sprang boldly up forward, the deadwood was built up aft and the long sloping sternpost began to define the outline of the ship. Meantime up in the loft Willard, with his helpers, was making up the sawn frames, doubling them and bolting them together, while preparations were going forward to steam the oak frames, bend them and set them up alternately with the sawn frames. Things were moving right merrily and then came the advent of placing the first plank. Weeks passed, and then bang went another plank. Another week, another plank. But they tell me that's how most boats are built.

You've gone too far to back out by this time, and they know it. You would not stop now though the heavens fall—they know that too.

Should I presume to criticise, which heaven forfend, I *might* say that they worked just a little erratically. I often wondered in my simple way, how it happened that when arriving unexpectedly and from the wrong side of the boat yard, I would see my boat standing neglected—shunned would not be too strong a word to use—the men passing and repassing without even seeing her. Here, I would think to myself, is where "Men's time and material used on Job 30" gets a little respite. However, strangely enough, within a few minutes of my unannounced arrival men would come running from every part of the yard and in a surprisingly short time the boat would be literally swarming with workmen working with an almost fanatical zeal.

I found the workmen a cheerful gang, friendly and interested. Neither Mr. Bones nor his immediate assistants seemed to be at all blind to each other's shortcomings, nor hesitated to proclaim them to the whole world when sure that the party criticised was in earshot. Willard would holler down to me from the gaunt carlins overhead: "If you'll just keep *him* away for a few days—" indicating Mr. Bones as he entered the loft—"maybe we can get this boat done." Mr. Bones' repartee was generally quite personal, but I never actually saw bloodshed during the building of my ship.

As a matter of fact everybody working on the boat seemed to be having a rather pleasant time. Of course, boat building is hard work. Some details mean lots of very hard work and highly skilled work too.

But the impression I got as an almost daily observer was that here was a good-tempered gang of men who knew how to do what they had to do, enjoyed the type of work they were doing and worked faithfully.

On my many trips to the boat yard I was often accompanied by a lady related to me by marriage, who was as interested in the building as I was. For her benefit Willard would mutter, apparently to himself but always within her hearing as he banged away with maul or hammer, "Bet the darned thing won't even float," or "Maybe I can make a boat out of this crate yet." Finally I made a deal with him that he was to accompany us on the first trip and carry in every pocket a chunk of lead sufficient to drown him if she sank—the lady in question thought highly of the idea.

The placing of the "whisky plank" is a milestone reached in the building of a boat and should always be duly celebrated. It is not for the prospective owner and his friends, but one confined to the men who are actually working on the boat. The whisky plank is the last plank to be fitted in place and then the hull is all closed in. A few bottles of Scotch placed at the men's disposal when the whistle blows on that particular day go a long way toward making the whisky plank quite a gay party and will bring the boat good luck. It also enables the prospective owner to convey to the men his appreciation of their work. Don't ever neglect the whisky plank.

It's funny how erratically a boat progresses while a-building. After the first few weeks I was almost prepared to see her in the water any time I was absent from the yard for more than a few days. But I outgrew that fear. For weeks the building seemed at a standstill. Then came weeks when the boat seemed to go backwards. I could visualize her returning to her component parts just as one would see a moving picture of the building run backwards, until nothing stood on the workshop floor but the keel. In such a picture it would have been funny to see Willard spring from a sitting position on the floor to the top staging of the boat, from which one day he took a tumble.

Nevertheless, work was going on somewhere. The "Job 30, Wages and Materials for week ending . . ." showed that. I would see two or three men putting partitions in the offices of the boat yard with "Job 30" time cards sticking out of their hip pockets. A gang of men watching somebody's boat being slowly run down into the water would automatically reach for their "Job 30" cards as the boat was towed away and they hunted other diversion. Mr. Bones' car was undoubtedly kept highly polished and his front lawn mowed by a man who kept a Job 30 time card close at hand. But went the work quickly or went the work slowly, or even when work apparently stopped entirely, one individual toiled like a spider up in the sail loft, spinning his web, night and day. And for his spinning he used marlin at so much a ball, and a wonderful web he must have spun, for he consumed hundreds of balls of marlin—at so much a ball. I made a rough computation and found that if all the marlin listed on Job 30 were stretched out end to end, one would have enough material to erect a suspension bridge between the earth and Mars that would carry a 3750-lb. truck with a 700-lb. load. As far as I could learn—and information was difficult to obtain—this marlin was all used serving the eye-splices in the rigging. Sometimes I think that building a boat makes one a bit suspicious. I like to feel, however, that Merle is reasonably truthful, and he told me that rigging takes a lot of marlin. Well, he proved it; but still—that was an awful lot of marlin.

The method of rendering bills for "Job 30, Wages and Material for week ending . . ." had also rather an original touch about it. Comes a huge envelope filled with long narrow typewritten forms. With a cold feeling around the heart the papers are riffled over. The first few sheets seem harmless enough—"2 sheets coarse sandpaper, 8c; 1 brass screw, 4½c; 1 sheet fine sandpaper, 4c;" and of course a few balls of marlin. The cold feeling around the heart disappears as the sheets disclose the use of hundreds of sheets of sandpaper, hundreds of brass screws, thousands of galvanized nails, all meticulously itemized and totalling perhaps seventeen dollars. Who wouldn't build a boat? But as the last sheet appears the cold feeling around the heart that has faded away is suddenly an icy clutch.—"Teak for deck, $7000.00. Total, $7017.42."— Just like that, no warning of any kind. Follows a long swim in a sea of doubt and fear.

Perhaps I should have omitted these mercenary matters—I didn't like them myself—and lest some potential boat builder may be discouraged, let him take comfort in the thought that the figures may be somewhat exaggerated. Two sheets of sandpaper, for instance, are nearer 6c than 8c. Nevertheless, Aspiring Boat Owner, there are lions in the path.

The building of the masts was another fascinating episode in the adventure, and I found my ignorance of modern mast building was ap-

palling. On the old *Pelican* the sturdy mast had stood thirty years' enthu-
siastic scraping and sandpapering and still remained huge in girth and
a tower of strength. We used backstays, but only for the exercise of setting them up. Every spring shavings were taken off that stick which
were nearly as thick as one of the sections of a modern mast. However,
there will be no more windy days spent up aloft in a bo's'n's chair scraping and sanding masts. That that pastime is gone never to return is very
evident. Should I ever consider it necessary to carry a spare mast, I shall
coil it like a wire shroud and stow it in the forepeak.

Masts are quite important and their construction is well worth studying and the art of spar construction today is not found everywhere. I
was helping one day to muzzle a sail on a large cutter (standing on the
extreme tip of the bowsprit), when the 125-ft. mast prepared to let go.
Luckily it changed its mind, but the heart beats I missed convinced me
that should I ever build a boat the masts should have my earnest attention. That same mast did carry out its threat at a later day, I believe, so
that it was not just spoofing us.

When building a boat another matter should always be borne in
mind: A boat is not complete when she is finished. Week after week
one hears the same old story—"Well, she's about finished now." "About
ready to shove her in." And the owner sticks even closer to the boat
yard. But still she lingers.

By the time the boat reached this stage of construction Dave had transferred most of the yard's equipment to the cabin of the boat, and apparently most of the repair work for the other boats in the yard was being
done there. Of course, it's more chummy working that way and it
saves sweeping the shop. All the shavings and odds and ends are neatly
piled in the bilges of the boat, together with old overalls, caps and as-
sorted garments of all descriptions. Which reminds me that there is
still an old hat way up aft that I must crawl up and get some day. I might
use it myself and thankfully.

Anyway, from the interior of my boat now comes a buzzing like a
hundred bee hives, and peering down through the hatch I see dozens
of men ostensibly putting the finishing touches to the interior, while
the planer drones and the circular saw screeches and the funny little
gadget that Dave cuts grooves with cries like a neglected child.

And still she lingers. But there is an end to all things and one fine day
we find her out of the shed in the sunshine. Her sticks are in and the

bowsprit on. No danger now of my sharpest fear being realized—for it's no secret that only the sharpest surveillance on my part prevented Bones from clandestinely building a clipper bow (which he loves) on to the stem of my boat.

Once out of the shed and in the open where one can gaze on her from all directions, one really sees for the first time the true lines of his ship. The false work and short range of vision in the shed did not allow a true perspective of the hull, but when I could walk around her and study her from every angle of approach, I realized that as far as I was concerned she was perfect.

The ceremony of launching drew near and called for thought and at this time I learned to my surprise that my wife's joyful anticipation of the day when the boat should slide gracefully and buoyantly into her natural element had become shot through with long black streaks of doubt and fear—with the fear that she might slap an eternal handicap on to the unsuspecting craft and launch her under a miasmic cloud of ever-threatening misfortune and dire mishappenings, by, forsooth, failing to break the promised bottle of champagne with the first swing and possibly pursuing the boat administering futile taps with the unbroken bottle until the belabored ship reached the water, unchristened and unnamed, a nautical nonentity.

" 'Tis well," said I upon learning the cause of her despondency, "that you are related by marriage to a man of considerable resource and intelligence, for this and so and in such a manner will I remove the barest possibility of the catastrophe you dread." And thereupon I discovered to her a method the adoption of which would ensure any sponsor carrying off her part of the affair with a grace and easy dignity that would cause the christened craft to slide into the water with a grateful grin of appreciation and a feeling of assurance as to her right to a place in the anchorage amidst her older and worldly wise sisters and to the name she legitimately bore.

Described briefly, my idea was to fasten a few feet of stout elastic to the stem of the boat and the other end to the bottle of champagne. The sponsor grasps the bottle firmly in her hands, backs off a few feet until the tautness of the rubber forbids further distance. Then, smiling graciously, she releases the bottle. With the crash of the broken bottle against the stem she hollers: "I christen thee *Mary Ann*," or whatever the name is, and the ceremony is over. I thought the idea pretty good, but my wife would have none of it. Moreover, my broken arm still being useless, she was able to demonstrate her ability to strike hard enough to break any bottle, so I let the matter pass.

About this time the sails arrived, and as many before me have done, I wondered at the uncanny skill with which a sailmaker manages to get six or seven hundred feet of canvas into a bag that would be a tight fit for a large pocket handkerchief. I am told it's the first thing a sailmaker is taught. Do you remember the picture of the Djinn escaping from the neck of a small bottle? Well, that is how the mainsail billowed out of its bag the first time it got free. My main cabin seemed too small for stowing a sail that came in a bag that a rather large lady might have carried on her wrist.

As I have already hinted, the day of the launching arrived. There she sat on her cradle with her tapering spars, her stainless steel shrouds and stays, her shining hull unmarred by smear or stain, and carven deep into her transom the name *Lavengro*. A gay string of signal flags borrowed from the *Prelude* stretch from bowsprit to mizzen boom and helped to complete the picture she made.

A small group of us climbed aboard and as the bottle crashed against the bronze sheave at her stem and the champagne spurted like a fountain, she slowly moved down towards the water. A hasty examination showed that the boat had suffered little damage from the violence of the sponsor's blow, the forthrightness of which well tested the stoutness of her stem-piece. I hope she never has to endure a shrewder blow than that christening konk.

A few exciting minutes and we felt her water borne as she floated off her cradle out into the channel, where she sat as gracefully as a drifting swan, her flotation lines well delineated by the film of oil which covered the water.

The show over, as far as the launching went, we towed her into a neighboring slip and made her fast for the night—her first night on the water—and how reluctantly we left her.

And how promptly we were back again the next morning. To our relief, she was still there, no evil having befallen her in spite of my fears.

Then commenced the last phase of her building. No one could accuse her of being a rush job, for she was launched the 220th day after her keel was set up on the floor, and I had enjoyed every day of it. What time the builders were not thinking up minor changes in her layout or gadgets of various kinds for her improvement, I would evolve a few ideas of my own, rush down to the yard and endeavor to sell them to the brain trust there. Generally my bright idea was disapproved of and I was often astonished at the unanimity of the whole gang, although once in a while they allowed me to hammer a small idea into the general design just to keep me encouraged. Personally I prefer a boat built with a little latitude as to minor design. If when the cabin is about roughed in it appears that a certain cupboard, shelf or whatnot, naturally belongs some place other than that shown on the blueprint, or even if it does not properly belong there but the owner wants it there, the sense of ownership is greatly enhanced by the change being made. It would often have puzzled an onlooker to tell for whom the boat was building when Bones, Dave, Willard and myself were all on hand at once, for each person seemed to have an equally strong proprietary interest in the boat and a desire to build her as well as she could be built. We affected to shut Bob out of our conclaves, as he had lost his heart to the beautifully lined Eight-Metre of Raymond Paige's which was rapidly approaching completion under the same roof, and we rather felt that his attitude toward us was just a little superior. The weight of eleven extra brass screws on the Eight-Metre gave him heartburn, while the *Lavengro* sported heavy bronze belaying pins on all pin rails and we all admired them.

Once the boat was in the water the really intense work was started. A deep path was worn from the shop to the slip as the workmen leisurely travelled back and forth laden down with a sheet of sandpaper or staggering along under a handful of cotton waste. It's a lot more fun working on a boat when she's in the water a couple of hundred yards away than when she's right in the shop.

Finally Bones said one day: "Better get her away from here, or we'll be working on her all summer."

So the next day the mattresses, which had been on hand for several months, were installed, bedding was brought aboard, all presented by interested members of the family, anchors and attendant gear stowed, also presented by interested members of the family, while the crockery and tableware were stowed by the donor. I had become an excellent collector and most of our cabin furnishings were donated. Dave had made us a present of a pair of beautifully grained rosewood knees worked out of a huge natural knee he had salvaged years ago from the wreck of a schooner off White's Point. Jim ("Jubilo") Dickson, my bitter sailing competitor, furnished the boat with a log of camphor wood salvaged from the old *Ning Po*, (built in 1756), from which Dave built a unique bookcase of delightful fragrance for the cabin, while that first evening we built a fire in the open fireplace—a present from the boat yard—from the piles of wood and shavings (presumably a present from the workmen) left in the bilges, the cupboards and under the bunks, just to see how the cabin looked with a fire in the fireplace.

The following morning, with pennant hoisted, we sailed from the yard and without a single adventure picked up our permanent mooring at the Los Angeles Yacht Club anchorage.

⚓

39

Tahiti or Bust

H. B. WARREN

Coop a yachtsman up in the harbor, and it is only natural that he should start dreaming of a cruise he will make—some day when the world is free. And the longer he is tied to the dock, the farther afield he will roam, in fancy. So it is that a spontaneous demand is growing for a Tahiti Race.

"But why Tahiti?" you may well ask. "Isn't everybody satisfied with the Honolulu Race? It's 2,200 miles and that ought to be long enough for anybody. The Hawaiian Islands are as beautiful as any spot on earth, and the people are so hospitable you are beholden to them for life."

All this is true. Yet it may be because of the many thrilling Honolulu Races that the demand for a Tahiti Race has arisen. Hundreds of men have now raced to Honolulu. The natural human craving is working on them to make a still longer cruise; and the biggest thrill of off-shore voyaging is having a brand new landfall rise out of the sea at the end of a long traverse.

Tahiti lovers claim vociferously that it is the most beautiful island in the world. The Pacific Coast boasts a large number of Tahitiphiles— people who have spent much time there and are eager to go back the minute the war ends. Burton Baldwin, widely known Los Angeles yachtsman, has been there twice, spending several months each time. He claims that the climate is even better than that of Southern California, especially on the windward coastline. There it was possible, before the war, to rent a grass hut and a Ford car, as well as to hire a native servant, all for about $30 a month. Food was cheap, for most of it was simply picked from nearby trees. Clothes consist of a *pareu* wound about the waist; so altogether life is simple and gay among the coconut palms and the good-natured inhabitants.

One does not have to make the long trek of 3600 miles to visit Tahiti alone. The Marquesas Islands lie between here and the Societies, while only a day's run farther are the atolls of the Tuamotus. Within a few days' sail of Tahiti itself are Moorea, Bora Bora and Raiatea. Farther South are lovely Manga-Reva and lonely Pitcairn. Some iron hearts may even make the first jump to Tahiti the beginning of a swing about the Pacific, or of a cruise around the world.

Many yachtsmen do not realize that a Tahiti Race was actually held, in 1925, when four San Francisco owners raced down there from the Golden Gate. The winner was the famous 107-foot schooner *Mariner*, the same boat that holds the elapsed time record for the Honolulu Race. Her time in the Tahiti Race was about 20 days, fast time under sail alone, especially when it is remembered that the Doldrums had to be crossed. This is still the world's longest yacht race.

Already two schools of thought have developed about the next Tahiti Race. The first group suggests that a cruise would be infinitely preferable. Their contentions, briefly, are:

1. The boats would undoubtedly vary greatly in speed, so that there might be as much as three weeks between the arrival of the first and last boats. Such events are far more fun when all the boats arrive nearly together. This can best be accomplished by setting a date when all hands will try to rendezvous, not at Papeete, but at Tai-O-Hai, on the Island of Nukahiva in the Marquesas. They point out that the Marquesas lie nearly dead to windward of Tahiti, and that this makes it much easier to stop there on the way down, which could not be done in a race. After a few days in the Marquesas, the next rendezvous would be somewhere in the Tuamotus, and the final jump would be to Papeete. This gives the slower boats the opportunity to start ahead on each leg, and with the faster ones overtaking them, many boats would sight each other and this wide stretch of ocean would not seem so empty.

2. This plan would enable each owner to use his engine whenever his speed fell below a certain amount, and this would shorten the voyage considerably. It might even save several days, since the Doldrums have to be crossed.

3. Using the motor will increase the safety while in the low-lying Tuamotus, because the time of arrival can be figured to come during daylight.

The racing men claim a cruise of this length would surely be unbearably dull and without the stimulus of competition the whole thing would have no value. From present indications this dissension will not wreck the whole idea. There are plenty of owners interested, so those who want to race will doubtless find all the competitors they need, while the cruising men will boast an equally big fleet.

Tahiti Races have been proposed and even scheduled several times since the first one. All fell through because of the time element. Very few owners or crews could take the necessary time away from business. In many harbors along the Pacific Coast owners are taking their time, but are steadily going over every bit of gear, preparing thoroughly against the great day that must come at long last—the day when sheets will be eased for the run down that 3600-mile roadway leading to the romantic South Seas.

It was altogether fitting that the San Francisco Yacht Club, oldest yachting body of our Pacific Coast, should sponsor the world's longest yacht race—the Tahiti Race of 1925. The idea was proposed by L. A. Norris at a luncheon of the club in November, 1924. At once, John C. Piver, Commodore of the S.F.Y.C., sensed the importance of the event and offered to enter his 85-foot schooner *Eloise.* His lead was followed by Dr. E. R. Parker, who thereupon pledged the entry of his 75-foot schooner *Idalia.* Norris, of course, entered the *Mariner,* which had established the record elapsed time for the Honolulu Race in 1923. Thus, there were three entries within a few minutes after the San Francisco Yacht Club had agreed to father it.

Commodore Piver quickly secured a promise from Sir Thomas Lipton of a suitable trophy, and all yacht clubs throughout the world were invited to send entries. Great interest was manifested by the clubs of the Pacific, but only one more entry was actually forthcoming, the 73-foot ketch *Shawnee,* owned by Mark Fontana of the Corinthian Yacht Club of Tiburon, California.

While there was much speculation about the proper course to sail in order to make the best time, there was little doubt in anybody's mind as to the eventual winner. L. A. Norris was probably the greatest Amateur off-shore sailor of his day. In 1910 he had the 89-foot schooner *Seafarer* built in Boston from Crowninshield's designs, and a shapely and able vessel she was. He took her across the Atlantic, through the Mediterranean and Red Seas, and across the Pacific to San Francisco. He reached here just in time for the Honolulu Race of 1912. *Seafarer* was undoubtedly faster on a boat-for-boat basis than the old *Lurline*, but the latter was commanded by Cap. Lew B. Harris, who had spent most of his life between Honolulu and the mainland. Consequently, *Lurline* sailed where the winds were strongest and won. Norris, however, kept the *Seafarer* very busy on offshore cruises until just after the end of the first world war, when he sold her and had the 106-foot schooner *Seaward* built from plans by Alden. This beautiful vessel, now owned by Cecil B. DeMille, was followed by the *Mariner,* a Burgess creation. Norris failed to win the Honolulu Race of 1923 with her, losing by a few hours to A. R. Pedder's *Diablo* on corrected time. But the *Mariner* set the elapsed time record of 11 days, 14 hours, 46 minutes, a record that has not yet been equalled. Norris brought many other boats to this coast, such fine vessels as the *Shawnee, Lloyd W. Berry, Radio and Navigator*. In all of them he cruised extensively in the West Indies, the South Seas and the Pacific Coast.

It was only natural then that the three other skippers should seek his advise as to the best course to follow. The Doldrums must be crossed on the way to Tahiti, and as their narrow portion shifts about with the calendar, the point at which they are crossed has a vital effect upon the length of the passage under sail. The pilot charts advised crossing the Doldrums some 300 miles eastward of the direct course. Norris took issue with this and said he expected to follow the great circle as closely as the winds would permit.

Mark Fontana, who knew Norris very well and understood his bluff, honest nature, followed his advice, which proved to be eminently correct. The other two boats elected to side with the pilot chart, and as a result took over twice as long to work through the Doldrums as the *Mariner* and the *Shawnee* did.

The start took place in the Golden Gate on June 10th, 1925, at 4:00 p.m. A healthy westerly was pouring through the Gate as the *Mariner* crossed in the lead. She dropped the others altogether too fast to suit Mark Fontana, skipper of the *Shawnee*. As soon as he could bear away to the southward therefore Mark set his ballooner. On observing this, Norris sent a man aloft to loose the foretopsail, which had been furled until that time. Both boats suffered the consequences. *Shawnee* lost her topmast and this hampered her for the rest of the race, which was mostly running and reaching. Her big ballooner and spinnaker did not set well from the stump of the topmast, and this fact seriously affected her showing.

The man on the *Mariner's* foremast was knocked overboard. Fortunately, he fell clear and was picked up by the cruiser *Missawit*. He was taken to a hospital, where he soon recovered. According to the rules, the failure to pick this man up disqualified the *Mariner,* but a quick-thinking committee, which was present on a power boat, immediately secured the consent of all skippers to waive this rule and permit the *Mariner* to remain in the race. This rule is designed for afternoon contests. In this case the cruiser was right at hand, and it would have been ridiculous and fool-hardy for the *Mariner* to attempt the rescue.

Norris went on to increase his lead steadily to the end. Both *Idalia* and *Eloise* outran the crippled *Shawnee*. The latter, however, did so much better in the Doldrums that she passed the *Eloise* to beat her by over a day, and all but caught the speedy *Idalia*.

The winds were amazingly steady throughout, while the weather was well-nigh perfect. All four boats sailed more than 3,800 miles through the water. The *Mariner's* elapsed time of 20 days, 11 hours, 32 minutes, is therefore only a little short of an average of 200 miles per day. Her worst day's run was 104 miles. This was in the Doldrums, and was much longer than one might expect to make at any time in that area with such a big heavy craft.

Eloise suffered the most hamper-difficulty of all. She dragged her centerboard outside the hull for over 2,000 miles. The pin holding it forward was corroded and carried away. The pennant was entirely too strong, so it was a long time before the crew could free the ship. It banged continuously against the hull, so it speaks volumes for the construction of the *Eloise* that she did not sink or even spring a leak. This mishap insured last place for her and came on top of constant trouble with the jaws of the main gaff. Even so, her time of some 29 days is not at all bad. *Idalia* was the first jib-header to take part in a long off-shore contest. The rig was comparatively new at the time. Dire predictions that the long mast would certainly carry away were not borne out in any way, nor did it give her any trouble.

The crews were royally entertained at Tahiti. It was Prohibition time, so the free-flowing champagne was a great luxury, and the tales of revelry have lost nothing in the telling during the years.

All four came home by way of Honolulu. As most of the passage from Tahiti to the Hawaiians is a reach, fast times were made on this leg. The fortunate few who took part in this great event still feel that it was the most interesting occurrence of their lives.

⚓

40

Professional Aid

ALFRED F. LOOMIS

I

THE eighth day of a transoceanic yacht race frowned on a sea rising in long, crest-tortured rollers, sinking in foam-flecked hollows. The sky, a gray ceiling of nimbus, darkened here and there over falling showers of rain; and the sea, reflecting the hue of the clouds, ineffectually attempted independence with its flashing whitecaps. The wind, ever the tormentor of sky and sea, pressed heavily from the west, ironically belying its force by the delicate tracery of its invisible fingers on the breasts of the waves.

At about the forty-fifth meridian of longitude and the fortieth parallel of latitude—an intersection discernible only to the human imagination—a small schooner of low freeboard drove across the tumbling confusion of the waves. There were men aboard her, and by that token the schooner was superior to the chaotic triune of wind, sky, and sea —she alone having definite form and pursuing a definite course. And these were men indeed, as could be told from the sail the schooner carried. It was not in the reefed mainsail that they asserted superiority. The two tucks in that expanse of canvas, bellying outboard to starboard, were, in fact, a concession to the pressure of the wind. But the number three spinnaker, its four-inch pole flexing like a willow wand, its thin canvas straining at the seams! That impertinent kite showed invincibility of mind.

And yet the men aboard the 54-foot over-all schooner, half of them sitting in the cockpit while the other half slept below, saw nothing magnificent in their audacious defiance. Those on watch—except the captain-helmsman, who occupied himself otherwise—looked steadily at the whipping spinnaker pole, at the frail triangle of cotton interposed before the rushing strength of the hard westerly. They would have said—had said, in fact—that if the Lord objected He could easily blow it away. Failing divine interference, if the spinnaker drove the schooner so fast that the sea sucked aboard her stern, the sail could readily be handed and passed below. While it preserved its integrity against the wind it added three knots to the schooner's speed and steadied her helm in the lifting, overtaking seas. There was appreciation in the eyes of the three idlers of the watch on deck, an amused quirking of the lips as they regarded the spinnaker and reflected that under ordinary circumstances a yacht like theirs would have been hove to. But they were racing.

The captain, who was also the owner and at the moment the helmsman of the schooner *Thetis,* looked only occasionally at the racing sail, and then only when the heave of a swell rolled the little ship to port and he wanted visual assurance that the spinnaker pole would not jab the wave crests. He steered with an automatic coordination of muscle and sense, a coordination so perfect that it almost defies division into its separate elements.

The helmsman's hands on the wheel, for instance, now lax and now suddenly white-knuckled, kept the schooner as true as might be on her easterly course—and it was the touch of the wheel which largely told him when to apply strength to right or left. So swift, so instinctive, was the reaction that the sensory impulses short-circuited direct to the muscles and even transcended instantaneity, to the end that for long periods of time the schooner hung immovable on her course, no more than a finger's strength sufficing on one spoke or another to keep her so.

Yet the little black-hulled schooner, presenting her stern to the onward drive of the rolling seas, was potentially able to outdo the strength of two men if more than an instant's inattention gave her charge. She could broach—that is the word of awful significance—and bury her nose and be pressed down by the weight of the wind in her sails while the sea threw high her stern and rolled her over. And then what of the men in her cockpit and those four below who had done their trick and had reposed their lives in the keeping of the helmsman?

But it was not by the delicacy of his strong hands alone that the captain steered. His eyes, clear, now snapping with enjoyment, now soft with content, watched intermittently the compass needle in the binnacle before him. That noiselessly oscillating magic of immovability gave the base course, the steering ideal. The hands, deceived by a groove in the sea when the yacht rode even-keeled and true, might have departed from the ideal by a point or more. But the needle, transmitting its immutability to the eyes and thence to the hands, brought her back again. Nor did the eyes linger in self-hypnosis on the compass card. They looked out ahead to see that the course was clear; they sought every minute the tell-tale whipping forward from the mainmast-head; they watched the tumble of the seas near and far, and ranged often the sails and rigging. Each glance of the clear blue eyes conveyed to the captain's brain a message of reassurance, each constituted an addition to his overflowing cup of timely knowledge.

With all going well, the sense of hearing was not called upon to aid the steersman's other senses. His ears picked up but let go the spasmodic conversation of his watchmates, and the overtone of the wind in the rigging. They were attuned only to the sibilant susurrus of the schooner's rush through the water, the rhythm of the waves overtaking.

In this art which the captain practised, the sense of touch informed him by another means. The wind, ruffling the short hairs of his neck, was a truer guide even than the masthead telltale. The eyes must impart many messages to the brain, but the skin has only to feel the direction of the wind. If the skin of the cheek as well as that at the back of the neck feels the draft, then the wind has shifted and some change must be made in steering. If, however, the cheek warms again, then it was only a temporary flaw and the course may remain the same.

Blending with all these sense impressions which made steering possible in that hard-pressed sea was the authority given by still another sense—the captain's sense of balance. At intervals the ship rose to a wave and for an instant hung. Then occurred a transition so slight as to be indefinable—so slight that not the compass card could detect it, not even the trained responsiveness of the hands on the wheel. But the helmsman's body as a whole felt the infinitesimal change in balance, and the anticipatory message was telegraphed to the wheel. Sliding off the crest of the wave, the yacht drove fast and true.

All this complicated human mechanism of steering was accomplished without fettering the imagination of the helmsman. His conscious

brain forged ahead to possible eventualities, reflected back to past experiences on such stormy days at sea. His judgment hovered in a state
of delicate equilibrium, ready to interpret an unusual sound in the
schooner's rigging or to seize a portent from the sea. On a moment
when the *Thetis* lifted high on a greedy, disappointed wave, he looked
ahead and saw a patch of weeds in the course. Instantly a ferment
started in his cup of knowledge.

II

The schooner had cruised for days the axis of the Stream, where
Gulf weed floats in long brown disrupted banners. She had plowed
through it, and her men, leaning over the side, had scooped up hand-
fuls of the growth to examine it for crustacean life. Gulf weed had been
a commonplace of the voyage. But, the Stream curving northeast while
the yacht continued east, its distinctive weed had thinned. This patch
ahead lacked the suggestion of buoyancy and mobility. Better, then,
not to sail through it, but to give it a berth and watch it as it went by.
At the next wave crest the patch was dead ahead and a hundred yards
away. Tenderly the helmsman altered course to starboard and prepared
to look overboard to port. The weed flashed by, and a wave in the
schooner's wake broke over it.

The helmsman spoke: "Boys, did you see that? The stump of a spar
with moss growing on it. Three feet in diameter and twenty feet long
—end on."

The three in the cockpit jumped up and looked astern. They sat
down. One of them spoke: "Hmph. Good thing you saw it, Charley.
It would have gone clean through us."

"Good thing it wasn't night," said another. "Bye-bye, *Thetis*."

The spell of silence having been broken, the captain, shifting slightly
on the wheel box, asked one of his shipmates for a cigarette. When it
had been thrust, ready lighted, between his lips, he puffed and offered
comment. "Good going, this. Wonder how the boys on the sloop *Alcazar* are making it?"

"I dare say they're carrying on," said the first speaker in the cockpit;
but added admiringly, "I never saw a boat pushed like this one,
Charley."

The captain shifted position again. "A grand rag, that small spinnaker. I don't see why it stays with us." Thus he disclaimed personal merit. Of his skill as a helmsman, no thought entered his consciousness. A clock struck in the cabin, its quick double notes faintly covering the rush of wind and water. "Read the log, somebody," continued Charley. "We're making knots."

One of the three rose and half climbed, half walked, around the helm to the low taffrail. He leaned over, his bare toes hooked over the mainsheet traveler, supporting and steadying himself on knees and elbows. Astern the white cotton log line spun dizzily and whipped the water in long, serpentine billows. The revolving wheel of the log stopped reluctantly as the sailor bent his hands from the wrists and brought the moisture-beaded dial into range of his vision. "Twelve point eight," said he. "That's—let me see—ten and a half miles since five. A tenth less than the previous hour. I wonder if the wind's letting up, Skipper."

The captain stole a second from his employment and cast a glance around the heavens. "Maybe," he conceded. "But as long as we're doing better than ten we'll carry on with this short rig. No use running risks."

"Oh, I wasn't criticizing!" exclaimed the sailor, steadying himself with a hand on the captain's shoulder as he stepped back into the cockpit. "If I were in command I'd have been hove to all night."

"Yes, you would," jeered one of his watchmates. "You'd be blowing away topsails, ten every hour."

The first sailor and the captain grinned. "You're a sail-carrying crew," observed the latter happily. "And look at the smile on the face of Chris."

The paid cook had emerged from the galley hatch and stood by the fore shrouds, reacquainting himself with the appearance of a stormy sky and sea. He looked aft and caught his employer's eye. "Where we go now?" he shouted. "To hell maybe?"

"Speak for yourself, Chris," returned the captain. "We're all pure aft here. How do you like ocean racing, Chris?"

The cook nodded his head in enjoyment and admiration. "You fellers sure know how to sail!" he exclaimed. "I'll get you a good hot breakfast."

A murmur of appreciation rose from the cockpit. "Good man, that," said one. "The first pro I've ever seen that wasn't sick or scared in an ocean race. But he positively likes it."

Chris, with one foot down his hatchway, took a look around. He pointed suddenly to northward. "Look!" he shouted. "Vessel in distress!"

"The *Alcazar,* I hope," said the captain, skeptically.

"No. Honest. A coal hooker or something. Mainmast gone and sales carried away. See the shirt in the fore rigging?"

Everybody jumped up, and one sprang to the weather main shrouds. "Yes," he confirmed. "Her hull's practically awash, and I see men waving from her quarter-deck. What do we do, Charley?"

"Get that spinnaker in quick. We'll have a look."

There was instant concerted action. The man in the rigging jumped down and ran to the lee pinrail, from which he upset the spinnaker halliard to the deck. Another jumped to the foremast, and the third, at the word of command, cast off the after spinnaker guy. The outer end of the spinnaker pole swept forward, spilling the wind out of the sail. The man by the foremast jumped the jaw of the pole clear and staggered aft with it. The man at the halliard cast off, but kept the line within the circle of his arms as he hauled down the shaking spinnaker and smothered it. As the racing sail came in, Chris, acting spontaneously, shot the staysail up.

"Snappy work, boys!" called the skipper. "Set the jib too, cast off the boom tackle, and then come aft on the main sheet."

III

For the moment interest in the discovered wreck was in abeyance, and even when the *Thetis* on her new course plunged toward it the crew were concerned with their own change of circumstance. Instead of flying smoothly (however dangerously) before the wind, they were now jammed hard upon it. Two men from the watch below, finding themselves thrown from their weather bunks to the cabin floor, came up, rubbing sleepy eyes. A vicious burst of spray doused them from head to waist, and they descended with howls of protest. The *Thetis* became a leaning, laboring thing, her decks and booms dripping and her bow rising and falling with a force that jarred. Instead of slipping quietly by, waves now broke against her port side, and the wind which whined evilly in the rigging threw the crests high.

"There's weight behind this breeze!" exclaimed Charley, whose helmsmanship was now concerned only with meeting the onrushing waves to best advantage. "Glad we haven't had this for the last eight days."

A shout from below preceded the eruption of four men from the cabin. They were all clad for heavy weather, and they scrambled to places in the crowded cockpit with expectancy in their faces.

"What's the big idea?" asked their leader, amateur mate of the amateur crew. He was large, his bulk accentuated by the close fit of the borrowed oilskin jacket into which he had thrust himself. The straining sleeves stopped short of his wrists and the button and buttonhole at the throat came not within three inches of meeting. With the first dash of spray his bare head glistened, while drops streamed from his rugged face and coursed unregarded down the strong column of his throat.

Charley glanced at this tower of strength affectionately. "Glad you came up, Hank," said he. "We've sighted a shipwrecked schooner, and we'll need your moral support."

"Not one of our compet— No, I see her. Golly, she *is* wrecked. Say, Charley, it's been blowing out here."

"And still is. We've got all we can stand under this rig, but I need both headsails for maneuvering."

"Right. What'll you do? Come up under the schooner's lee?"

"Yes, but they'll have to jump. We can't go alongside."

"We-ll," the mate drawled in disagreement, and then changed his mind. "I guess you're right. We've still got the race to think of. Hope they can swim."

Each corkscrew heave of the *Thetis* brought her nearer to the wreck, which was now seen to be heeled to an appalling angle. Five men clung to the weather rail of the slanting poop deck, and their calls for help came thinly down the wind. Charley, on his feet now, sized up the situation. He knew little of merchant schooners and could not guess the life expectancy of this one. She might float for an hour or a month. His problem was how to approach her. There might be—there was— wreckage to leeward. He must not go too close. And yet he must not expect her crew to swim far. Perhaps they were on the edge of exhaustion. Nor could he throw his own vessel into the wind and let her lie there indefinitely, her sails spilling. Slatting about, they would blow to pieces.

"Boys," he suddenly said, "her stern's pretty much up in the wind, and we'll pass under the bow and come about to weather. I'll luff past as close as I dare, and we'll tell 'em what to do as we go by. Then we'll have to work fast. As soon as I get room I'll run off, jibe over—"

"Jibe, in this?" someone asked incredulously.

"The boom will be hauled flat. She'll stand it. We'll jibe, shoot up, and lose headway abreast her stern. Hank, you tend jib sheets; Chris, stand by to back the jumbo; and the rest of you heave lines and haul the men aboard. Somebody fetch the megaphone now, and stand by to come about."

They passed close to leeward of the wreck. Her mainmast, with its smaller spars and shreds of sail trailed in the water, still held by the starboard shrouds, which on the instant snatched the sticks back to punch hollowly against her. Of her decks all but the poop and forecastle were under water, and the sea tumbled over her weather bulwark to surge convulsively over her waist. The foremast and the bowsprit, still upthrust, seemed to lean despairingly from the wind's blast. A sound of the groaning of tortured wood and wire came to the *Thetis* as she plunged by.

Now the *Thetis* tacked to weather of the wreck, and the crew saw how her main chain plates had been torn away, opening up her port side and letting the mast go by the board. This side was high out of water—at least, as the breaking waves fell away from it—and all the opened seams of its black planks wept rivulets.

A leaping sea threw the *Thetis* bodily so that scarce twenty feet of open water kept her from the sullen hulk, and a voice which carried upwind arose from her poop. "Sheer off, you fools! You can't do anything to weather of us!"

Charley smiled as his crew watched him anxiously. He raised his megaphone. "Pipe down and take instructions. Can you all swim?"

"All but one. What's the matter with your dory?"

"It won't live in this sea. I'll round up to loo'ard. Jump as you see your chance. We can handle you all at once?"

Hank added a postscript. "Here's a lifebuoy with a rope. Give it to the one that can't swim." He swung his powerful arm and a white ring bounced over the ship's rail. A grasping hand caught the attached rope and Hank let go its end. The *Thetis* passed astern.

Instantly Charley brought up his helm and paid her head off. Her jibs, from curved, straining boards, became gently shaking cloth. The spray dropped and the motion eased, but now the mainsail, feeling the wind on its leech, began jumping and pulling intermittently at its taut sheet. "Jibe oh!" cried the captain. "Weather jib sheet! Hold on!" With a sudden shift from port to starboard the main boom swept rebelliously through its narrow arc, and for the instant that the *Thetis* swung broadside to the wind she lay over on her starboard beam ends. Then, rudder and pressure of wind assisting her, she whisked around, presented her bowsprit to the eye of the wind, righted, and lost headway. She lay where her captain wanted her, no more than a long jump from the hulk, smooth water between.

"Your only chance, men!" he shouted through his megaphone. "I might crock up next time."

They slid, scrambled, and fell down the sloping deck and plunged, heads thrown back, into the water. Three swam independently and reached the *Thetis's* side in ten strokes and were hauled aboard. The fourth wore the white life ring beneath his arms, and the fifth paddled with the end of the rope in his fist. He passed it up to reaching hands, and turned back to his helpless comrade. But impatient voices restrained him. "We've got him, all right. Come within reach. We can't lie here all day."

To these persuasions Charley added his. "We're gathering sternway, boys. If we get meshed in those spars and rigging we'll never get out. Now! I've got to let her fill away. Back that jumbo, Chris."

"Heave ho!" cried Hank, hauling in the port jib sheet, but watching the rescue operations over his shoulder. "There are five aboard us, Charley. Is that all, Cap?"

"That's all—and damn glad to be here. Where are you bound?"

"Never mind that now," answered Charley. "I'm jibing again." His heart thumped with the exultation of a dangerous job well done. His eyes shone. "Somebody write up the log and get their names and facts. Oh, and the patent log. Did that foul anything?"

"No. It was taken in."

"Fine. Stream it, and—jibe oh!"

Again the close-hauled mainsail thundered over, and now, as its sheet was slacked out, the *Thetis* resumed the long rushing roar of her former gait. The derelict dwindled rapidly over her port quarter.

"I guess you can set that spinnaker again. No. Wind's moderated a lot. Make it the size larger. . . . Well, men, you're welcome to what hospitality we have."

IV

To the crew of the *Thetis* these five shipwrecked mariners who lay exhausted on the schooner's deck were Titans of the deep. They belonged to that unfathomable, almost mythical order of beings who keep to the sea in all seasons to wrest a scanty living from it; who, with inadequate equipment and in insufficient numbers, drive ponderous schooners through winter gales, and arrive, overdue, unconscious of their heroism. These five, who had suffered shipwreck and stared death in the face, who had accepted rescue without visible emotion, were objects of special admiration. Under their eyes, and particularly under the eyes of their captain, who had shown his contempt of amateurs in the moment before his rescue, the Thetans must sail with every ounce of smartness at their command.

As was to be expected, the story of the mariners' privations was simply told. In a hard blow seven days previously the schooner *Maribella's* cargo had shifted. To top that, the main chain plates had pulled out of timbers which had long been rotting, and the mast had given way. A week of pumping had been in vain. The stores were wet and the fresh water was gone. The end had been in sight when the *Thetis* came up. Luck had been with them, and now where were they bound?

Hank brought brandy, and Chris fresh water and biscuits, and promised hot food within the hour. Feeling the stimulant, the shipwrecked ones sat up and looked about them in amazement.

Their captain, who gave his name as Duggan, voiced their wonder. "What the hell is this little peanut shell doing here with no harbor to run to? Racing? Where? To England from New York? What for? For the *sport* of it? Could anybody be so crazy as to look for sport in midocean in a thing like this?"

These questions prodded the pride of the Thetans. Their schooner was a staunch little ship, designed especially for ocean cruising. They raced her because there was no sport like it—no other sport in which man pitted his wits against the elements while in competition with his fellows.

It was Duggan's opinion, candidly expressed, that if they wanted to live to race another day they'd better be jogging along under foresail and wung-out jumbo.

"But what about the other birds?" asked Charley, who had been relieved of the wheel. "There are four of them back there who won't be jogging along. They'll be carrying on."

"What! More of them doing the same?" asked Duggan, his wonderment increasing. "Well, if there were any professional seamen in the lot they'd be riding easy under squaresails."

Professionals. The Thetans had small opinion of such as ship aboard yachts, looking for soft berths and generally finding them; and these shipwrecked mariners had been excluded from that category. But here was the classification out of Duggan's own mouth—if they had the wisdom of professionals they'd be playing safe under squaresails.

"If any of our adversaries are carrying squaresails," said Charley, "you don't have to ask why they're behind us. This number two spinnaker of ours makes us know we're racing."

"So that's what you call that balloon, eh? I was wondering what it was. Looks to me like a man-killer." Duggan cast his glance aloft. "Look at your spars buckling. And look at that damn slender preventer stay. It isn't heavy enough to seize a clew to a boom, and if that lets go you'll be like we were a week ago."

"That's a chance we have to take," said Charley; "and I hope you won't feel you've jumped out of the frying pan into the fire."

"Who, me? Race your fool heads off for all of me. But suppose the ship does break up beneath your feet and you have to take to the small boats. Where'd you be in that dory, even without the five of us?"

The crew of the *Thetis,* attending to this conversation with interest and a sense of disillusionment, glanced at the dory lashed bottom up on deck and grinned. It was intended for ferrying men to shore in quiet harbors. Five was its maximum capacity in still water. There were now fourteen souls aboard. The dory situation was one of the inherent humors of ocean racing. It never had seemed more laughable than at the present moment.

"We'd have to swim ashore," said Charley; and the conversation lapsed.

Around midday, by which time the rescued mariners had fed and had fallen into a heavy sleep below decks, the wind moderated still more, and changes were made in the schooner's sail spread. The reefs in the main were shaken out and the whole sail hoisted. The spinnaker was taken in. The balloon jib and balloon staysail were set, and the course was slightly altered so that these swollen acres of canvas would fill and draw to top advantage. By these changes the schooner's speed was maintained despite the softened wind.

As Duggan came topside in the afternoon his eye lighted to see blue between the scurrying wefts of cloud. But his square, unshaven jaw dropped as he looked forward and observed the schooner's mountain of canvas. Speechless, he walked gingerly to the foremast and with his tough fingers felt the texture of the balloon staysail. It was thin, like the cloth of a much-laundered shirt. He returned aft and sat down. Like a man in the zoo he inspected one by one those of the Thetans who were on deck, seeming to see rare specimens in which indications of rampant madness were all too evident. But he remained silent, neither displaying interest in the badinage which flashed back and forth between the lighthearted watchmates nor offering to help them in the minor details of ship's work which engaged their hands.

Three of Duggan's shipmates dribbled up, refreshed, and grouped themselves compactly near him. They were clad in an odd assortment of flannel trousers and varsity sweaters with the initials turned in. In response to questions they declared that they had never felt better; but they too seemed disinclined to talk or mingle with their rescuers. They exchanged words among themselves, but these were monosyllabic. They touched cleats and rope ends and such other small objects as lay within their reach—touched them wonderingly, as one will a baby's hand, or a tiny bird's egg.

At supper time the commander of this incomprehensible craft came up from a berth which he had fashioned for himself on the cabin floor. He inspected carefully the stand and trim of the sails and climbed aloft to look for signs of chafe. Satisfied, he came down and for some minutes watched in silence the run of the sea and the appearance of clouds and westering sun. At length he gave the result of his deliberations.

"I think we'll have a good night, eh, Hank? Certainly no reason for shortening before sundown."

"Everything's as slick as hair oil. We batted off 240 between afternoon sights, yesterday and to-day. And that's going."

Captain Duggan stirred and spoke. " 'Scuse, me, Cap'n, but you ain't thinking of carrying this light stuff all night, are you?"

"Why, yes. Every mile we make in this westerly weather is good for two miles at the other end. Play your luck while it lasts, or it won't last."

"I was just wondering. S'pose there was some other derelict like the *Maribella* on your course at night. What then?"

Charley shrugged his shoulders. "I've also heard," said he, "of icebergs, and ships struck by meteors. We take those chances."

"At least you keep a proper lookout?"

"I've been thinking of that. You noticed, I suppose, that we have places in the two cabins for only six to sleep at one time. Your cook has gone forward to help Chris, and there's a spare berth for him in the fo'c's'l. So that leaves just a dozen of us aft. Now, I don't want to make you work, as we have a full crew without you; but I'm afraid you'll have to stand watches with the rest of us, so there'll be room to sleep. If you and your men care to do lookout duty, it would be a first-rate solution of the difficulty."

"That's fine. Men, we'll keep regular watch order, and I'll stand with the captain here. And, Cap, don't think we don't want to work. Anything we can do, or any advice my mate and I can give, we'll be glad to."

V

No doubt it was the memory of his almost fatal shipwreck which warped Duggan's weather judgment in the continuing days of fine westerly weather. This, and his deep-rooted conviction that a yacht less than sixty feet long was a rich man's toy, fit only for harbor sailing. The advice which he contributed with less and less reserve was always on the side of caution. Fair-weather clouds, when robbed of the lingering luminosity of the setting sun, became the forerunners of black squalls. Minor fluctuations of the barometer aroused his concern.

Once, calling upon his years of experience to back up his dicta, Duggan persuaded Charley to take in his kites on the advent of a midnight squall. But his acceptability as a weather prophet terminated when with the lapse of two hours of expectant waiting nothing happened.

Duggan's men, ever suspicious of the amateur's sailing ability, but faithful to their duties as lookouts, met their Waterloo on the day when, the fine weather ending, the *Thetis* crashed into an easterly. This was in the Chops of the Channel, where the ocean shoals and the waves are steep. The Thetans could and did make allowances, for they had been unmercifully shaken up the day after the start. They knew, too, that a man who is immune in big ships or even in ships of moderate size may succumb to the violence of a small yacht lying on her ear in a short head sea.

So on this revealing occasion the Thetans said nothing, and did not even exchange meaning glances among themselves. But the distressed mariners, more distressed now than they had been in a lifetime of sailing, dropped their heads in mortification. Lookout duty might have seemed to them a supererogation when each tortured, sea-whipped lurch of the frail *Thetis* promised to be her last. They huddled wet and miserable during their tours on deck, and one of them expressed the sentiments of all when he said, "We knew we were going to drown on the *Maribella* schooner, but this damn being half drowned and half bounced to death is what gets me."

When the strong clear easterly gave way to a thick south-westerly and the *Thetis* once more laid her course, her captain showed first signs of worriment. He was now, after more than two weeks of unlimited sea room, running fast on a lee shore, and a reliable fix was as important as the need for making every minute count. But here luck intervened. At noon the sun showed itself long enough for an accurate shot for latitude, and two hours later two coasters crossed the *Thetis*'s bow— one bound north and the other south.

At sight of them the worried frown left Charley's face. "Boys," said he, bringing a folded chart up on deck, "here's where we are. Latitude by observation, forty-nine, fifty-two; longitude, a line drawn close to westward of Wolf Rock. See? That's where those coasters are going —one south out of the Irish Sea, having rounded the Longships and given the Wolf a berth, and the other on the reverse course." He gave the helmsman a steering order that, allowing for tides, would take them close past the Lizard.

But Duggan interposed his last objection. "I want to say, Cap, that in the last nine days I've changed my mind about yachts and gentlemen sailors. I take off my hat to you for making a schooner go. But going it blind on a day as thick as this ain't seamanship."

"How do you mean, 'going it blind'?" asked Charley. "My latitude was good; and what could be better than longitude gained from those two coasters? Don't they know their way?"

"Yes, they know it; but you don't. They might be going anywhere but where you say."

"But where? Would they be running onto the rocks of the Scillies? Or full bore into Mount's Bay? And we can't be as far east as Plymouth. You'll find I'm right, Duggan. I've cruised this region."

"That's all right, but if this were my ship coming onto a foreign coast I'd feel my way. You'll pile her up, and then what will the underwriters say? Were you taking it easy? Did you run a line of soundings?"

"I'm not insured. So what do you say, boys?" And Charley put it up to his men.

"I say I'm with you until the keel rises up through the deck," said Hank. "That bad spell of easterly weather let the sloop *Alcazar* and probably a couple other windward workers slip through us. At least we don't want to finish last."

So the final hundred miles were run in an atmosphere brittle as icicles. The Thetans felt intuitively that if ever they had held the ascendancy over their rivals they had lost it in the head winds. The Maribellans knew that they would yet have to swim for their lives, holding to the gunwales of that ridiculous dory. And the quartering sea roared, and invisible steamers bound down Channel shaved them as darkness came in, and every man jack stayed up to see the finish.

The siren of the Lizard boomed too close as they flashed by it in a thin fog. But it sounded when and where the captain of the *Thetis* wanted it. And four hours later—but it seemed like fifteen minutes —they clocked their time of rounding Plymouth breakwater and brought up in the anchorage. No committee came to greet them, and until morning there was no way of telling whether they were first boat in or last. This gave to the transoceanic its final fillip of excitement.

There were people at home fully as anxious for news of the finish of the race as the crew of the *Thetis,* and they had only another day to wait for it. It ran, from the facile pen of a shore correspondent:

"At 12:15 Monday morning the yacht *Thetis* won the transoceanic race in the remarkable time of seventeen days and seven hours, setting up a record for small yachts that may stand for many years. She defeated her nearest competitor, the sloop *Alcazar,* by twenty-three hours and forty minutes. On the harrowing voyage the *Thetis* figured in the thrilling rescue of the captain and four men of the merchant schooner *Maribella,* abandoned in mid-Atlantic. While there is no inclination here to belittle the sterling performance of the amateur crew of the *Thetis,* it is believed in shipping circles that they could not have won such an overwhelming victory without the superior ability of the professional seamen from the *Maribella.* If this is true, the race must take its place among the stirring romances of the sea. In return for their lives, Captain Thomas Duggan and his men from the *Maribella* showed the amateur sailors the way to the winning post. . . ."

There was more in this vein, but a little should suffice.

⚓

41

The Loss of the "Leiv Eiriksson"

HERBERT L. STONE

On July 4th, 1924, two members of The Cruising Club of America, William Washburn Nutting, the first Commodore of the Club, and Arthur Sturgis Hildebrand, left Bergen, Norway, in a small cutter rigged double-ender, or skoite, of Colin Archer design, on a voyage to America with the intention of following for part of the way the route the Vikings took a thousand years ago on their voyages to Iceland, Greenland and Labrador. Nutting was in command of the expedition, and besides Hildebrand, the crew consisted of John O. Todahl of New York, and Bj. Fleischer, a Norwegian yachtsman.

The little ship was named, very appropriately, *Leiv Eiriksson*. She was 42½ feet over all, strongly built and well equipped for the long passage, which was to include, if ice conditions permitted, a call at the east coast of Greenland, which is rarely visited.

After calling at the Faroe Islands the *Eiriksson* reached Reykjavik, Iceland, on July 25th, without mishap. Letters written by Nutting and received subsequent to the departure from Reykjavik gave the date of their proposed departure for Greenland as August 10th. In these letters he said that ice was reported to be very heavy off the southeastern coast of Greenland that they would not attempt to make the settlement there but would coast along the edge of the ice to the southwest coast of Greenland, and go from there to Battle Harbor, Labrador, where they expected to arrive prior to September 15th.

Up to October 1st no apprehension was felt for the safety of the party. But receiving no further report from the ship, or no word having been received of her arrival in Greenland, action was taken early in October in an endeavor to locate the yacht. Through the courtesy of

Reprinted by permission of *The Cruising Club*.

Harry Cox, a cable, via Denmark, was transmitted through private wireless stations to Greenland, asking whether the *Eiriksson* had made port there. The reply was that she had not. Requests were also made of Western Union cable stations in Nova Scotia and Newfoundland to report any information obtainable through shipping and amateur wireless stations. Through Harry Greening a request was forwarded to all Hudson's Bay stations on the Labrador coast asking that they be on the lookout for the yacht and her crew.

In the meantime the "Norgensposten," a Norwegian newspaper of Brooklyn, received from their Norwegian correspondent a report that a letter had been received from Mr. Fleischer, stating that the *Leiv Eiriksson* had made port at Julianehaab, Greenland, and would leave for Battle Harbor, Labrador, on September 8th. In compliance with our request, confirmation of this report was obtained by cable. At the same time a further cable was received from Greenland confirming the departure of the yacht on September 8th, for Battle Harbor.

On the arrival of Donald B. McMillan from the Arctic and the Labrador about the middle of October, he was communicated with, and reported he had no word of the *Eiriksson* or her crew. He also advised us as to weather conditions along the route of the *Eiriksson* from September 8th to 17th, which were most unfavorable.

So, on October 24th, a request was made to the U. S. Navy Department to lend their assistance through the immediate dispatch of an ice patrol vessel or revenue cutter to search for the missing yacht. A similar telegram was also addressed to the Secretary of State, asking that he enlist the aid of the Canadian, Danish and Norwegian Governments. Prompt advices were received to the effect that the State Department had complied with the request and that the Norwegian Government had already notified them, through the United States Minister, that immediate action would be taken. The Danish Government also confirmed the sailing of the *Eiriksson* from Julianehaab, September 8th. On October 30th and 31st, Henry A. Wise Wood telegraphed to the Navy Department and to the President of the United States restating the facts of the case, and pointing out the need for immediate action.

These messages brought a prompt reply on October 31st that the U. S. cruiser *Trenton* had been directed to proceed to sea to search between Greenland and Labrador for the missing ship, after getting all the information available from officers of the Club.

In the conference between Captain Kalbfus of the *Trenton* and Messrs. Wise Wood and Kattenhorn it was decided that the search should be made between Meridians 27 and 45 West and Latitudes 54 and 59 North. Captain Kalbfus pointed out that they should have a navigator aboard familiar with Arctic conditions and Captain Bob Bartlett was asked to go. Without thought for lecture engagements and general disruption of his personal affairs, he caught the first train for New York, to confer with Captain Kalbfus and Wise Wood, and sailed in the *Trenton* when she left on November 3rd.

At the instance of Captain Kalbfus advertisements were also inserted in the New York, Boston, Halifax and London papers, asking information of the *Leiv Eiriksson,* with the hope that data might be obtained from some vessel which might have sighted her.

The search of the *Trenton* occupied 12 days. In order to understand the difficulties entailed in the search, excerpts from the official reports of Captain Kalbfus, of the *Trenton,* are appended. One of these is his estimate of the situation, which governed him in the area to be searched, and the other is his report on the actual search itself, which covered some 4,100 miles of sea and was carried out under unfavorable conditions, the time of year being such that the sea was rough, preventing the use of scouting planes, and the hours of daylight were short:

Estimate of the Situation

"The *Leiv Eiriksson,* a 42-foot skoite, double-ended 14½ feet beam, 6 feet draft, equipped with a two-cylinder Kelvin motor, and cutter rigged with hollow boom and gaff, left Julianehaab, Greenland, on September 8th, 1924, for Battle Harbor, Labrador. Four men were on board. She was not equipped with radio except a small (amateur) receiving set. The date and fact of departure have been confirmed through three separate sources. It was the plan to proceed to Baddeck, Nova Scotia, if it was not possible to make Battle Harbor. The vessel was seen in Iceland, on August 8th, by the officers of the U. S. S. *Richmond* and she actually came alongside the *Richmond*. Provisions were offered the *Leiv Eiriksson*. She took but few items as she was well stocked at the time.

"It is 600 miles in a direct line from Julianehaab to Battle Harbor. The nearest point on the coast of Labrador is 500 miles from Julianehaab and by first making the Labrador coast the trip to Battle Harbor would cover 700 miles. It is definitely known that she had not arrived in Battle Harbor on November 2nd, 55 days after her departure.

"In a complex problem of this nature, where the exact information consists only in the date of departure, the destination, the course (within limits), and the expected time of arrival, and where a single ship is conducting the search, the area of which is at a considerable distance from her base, it is necessary to examine into all aspects of the case and to base the decision on the most probable assumptions. Only in this way can the search be conducted profitably. The area that can be examined is necessarily limited in extent although the possible area is almost limitless. In considering the various assumptions the least likely must be rejected. (The assumption that seemed most likely to Captain Kalbfus was that the vessel became disabled on her course to Battle Harbor, but more than 100 miles from either Greenland or Labrador).

"The assumption being, therefore, that the vessel became disabled sometime between September 9th and 16th, the problem becomes that of calculating and estimating her probable location about two months later. It must be assumed that, if afloat, she is totally disabled and drifting, as otherwise she would have had time to make Newfoundland. Practically two months have elapsed since she was last seen. The Labrador current flows in a southeasternly direction at an average rate of 11-12 knots per day. This is the minimum average drift that can be assumed; any additional influence, such as strong northwest winds, would augment this drift. The conclusion is reached, therefore, that a line parallel to the line joining Julianehaab and Battle Harbor, and distant from it a number of miles equal to the elapsed time, multiplied by the minimum average drift, represents the northwestern limit of profitable search.

"To determine the most likely sea area there is nothing to base the calculation on beyond the known fact that the Arctic current flows out from between Greenland and Labrador at an average rate of about 11-12 miles per day in a general direction of about 150°, and the further fact that the prevailing westerly and northwesterly winds would augment the rate of drift of a floating object. The well known position of icebergs substantiates this. These considerations determine the Commanding Officer to make the following:

"*Decisions:* 1. To search from the flank on a line distant from the probable line of the *Leiv Eiriksson's* course, approximately 12 miles per day multiplied by the number of days of elapsed time from her probable successive daily positions on her course.

"2. Noting that the great steamship lanes are approximately parallel to this line, to utilize this fact by considering the transatlantic steamships as additional scouts which will cover the estimated 'enemy's' speeds of about 13, 14 and 15 miles (drift) per day.

"3. To retire across the steamship lanes in a southeasterly direction and search from the flank on a line approximately parallel to the first line and covering a drift of about 16 miles per day.

The Actual Search for the "Leiv Eiriksson"

"The *Trenton* sailed from New York the forenoon of November 3rd, speed 18, the general plan being fixed but the exact details remaining to be worked out. The positive opinion had, however, been reached that the *Leiv Eiriksson,* if afloat, could not possibly be less than 600 miles from Battle Harbor in a general southeasterly direction at the time the *Trenton* sailed. This fixed the general area in which the *Trenton* should search and the opinion of the Commanding Officer in this regard was unanimously concurred in by all of his senior officers, as well as by Captain Bartlett. Course was accordingly laid for a point south of the Great Bank of Newfoundland, it having been decided that nothing would be gained by crossing the Banks in view of the large percentage of days of fog at the time of the year and the fact that the Banks would not be a profitable searching area because of the large number of fishermen frequenting the Banks, which fact would have made it likely that the *Leiv Eiriksson,* if on the Banks, would have been picked up by these fishermen.

"As soon as the *Trenton* cleared Ambrose Channel the greatest possible amount of study, in detail, was given to the problem. Conferences of the senior officers, to which Captain Bartlett was invited, were almost continuous. It was recognized by all concerned that the task would ordinarily be beyond the capabilities of a single scout or even, in view of the elapsed time, the capabilities of a larger number of scouts. This however, had no influence upon the determination of the officers of the *Trenton* to search thoroughly the most likely sea area in accordance with the plan. It was realized that there would be no assistance such as smoke to aid the search, and that the band of search on either side of the scout was necessarily narrow, because of the size of the vessel being searched for. The daylight hours were short, although the moon was full during part of the period. The search was actually adversely affected by the fact that the seas were frequently heavy, which reduced the width of the band of search, and the fact that the hydro-aeroplanes could not be used because of the strong winds and heavy seas.

"Snow flurries were encountered in the vicinity of the Great Bank of Newfoundland and the weather during the period of search proper was bad. The wind frequently reached the force of 9, coming in great puffs when it did so. The vessel behaved splendidly throughout, although she was not driven at high speed into the heavy seas.

"Inspection of the reports of the weather from Belle Isle since September 8th, and the opinion of Captain Bartlett, indicate that the weather encountered by the *Trenton* was normal for this time of the year. While it is known that a small boat can remain afloat in almost miraculous fashion in a heavy sea, it did not seem to the officers of the *Trenton* that a 40-ft. craft could possibly live in the seas through which the *Trenton* passed, unless the personnel on board were skilled and alert and the small craft in a normal operating condition, which would scarcely be the case after such a lapse of time.

"The fact that the *Trenton* was searching was broadcast to all vessels with the request for prompt information if anything resembling the *Liev Eiriksson* was sighted. On November 6th a report was received from the S. S. *Aroli Amendi* to the effect that at 9:00 a.m., on November 4th, that vessel had passed a floating obstruction in Lat. 40 —36 North, Long. 57—29 West. The captain of the *Aroli Amendi* was requested to furnish further information and in response reported that the obstruction was about 40 feet long and 14 feet wide, apparently cutter wreckage. These dimensions agreed exactly with those of the *Leiv Eiriksson* and accordingly the *Trenton,* being outward bound at the time, this information was broadcasted to all vessels with request for additional information. In view of the fact that this wreckage was only about one foot out of water, with no signs of life on board, the Commanding Officer decided to continue his search to the northward and eastward and to make the wreckage upon the return. This was done, the probable drift in the meantime was calculated and the probable resulting spot as well as the original spot were examined, but no trace was seen. As the weather at this particular time was fine and the sea smooth, it is probable that the wreckage had gone down, and this opinion is substantiated by the fact that it has now since been reported.

"During the nights of the search the searchlights were kept playing and many additional lookouts were kept stationed. The radio was constantly manned.

"Upon the return of the *Trenton* to the southwest end of the second search line, it was decided that to examine further greater probable drift speeds would not be profitable. One reason for this decision was that higher drift speeds than those examined by the *Trenton* would carry the vessel close to the Azores and the low powered steamer lanes, the likelihood being that, if the *Leiv Eiriksson* had reached this area with those on board alive, she could probably make the Azores. Accordingly the *Trenton* proceeded to examine the area in which the floating obstruction, referred to above, was reported, and then returned to Tompkinsville for fuel, having reported to the Navy Department the fact that the search had been conducted but without success. The distance steamed was 4,092.1 miles.

"From the beginning the Commanding Officer has felt a keen interest in the probable fate of these intrepid Americans who undertook a voyage of this nature in a small boat and, obviously, too late in the year. Throughout the search he refused to entertain the idea that the crew of the *Leiv Eiriksson* had met their death. So many things could have happened that it is impossible to make a true guess as to what did happen. The days of anxiety and hope of a chance encounter on the part of the officers of the *Trenton* during the progress of the search led to the firm conviction by all that the *Leiv Eiriksson* is not afloat on the sea with her personnel alive."

—*E. C. Kalbfus*

Through the efforts of Harry B. Greening, the Hudson's Bay Co., during the winter of 1924–25, instructed all of the Company's posts in Baffin Land to be on the lookout for traces of the ship or party, and if found, to harbor them over the winter and get them across to Labrador during the summer of 1925. They discovered no trace of the party.

During the summer of 1925 an expedition was sent out from Norway, with the consent of the Danish Government, and a search was made of the southwest coast of Greenland, but no trace of the party and no wreckage was found which would give any indication of its fate.

What might have happened is purely conjecture. No one knows. Probably no one ever will know. Captain Bartlett believed they hit a "growler" or floating ice, near the Greenland Coast, and went down. It has been reported that at Julianehaab Nutting was given two carrier pigeons to release if the ship got in trouble. If this is so, the fact that these birds never returned indicates that the end came so quickly that there was no time to set them free. It is some comfort to think that this was likely, and that the *Leiv Eiriksson's* crew were spared long suffering and hardship.

⚓

42

Notes on Ocean Racing

THOMAS FLEMING DAY

In ocean voyaging and ocean racing fortune has sometimes much to do with success, but I have found by constant vigilance and skilful maneuvering you can oft-times anticipate her gifts, and thereby gain a double advantage over a slower and less lucky adversary. She will often by a happy deliverance save you from the calamities of an error, correcting a mistake of judgment by an unexpected, opportune shift of wind that changes your position from one of loss to one of advantage. But it does not do to solicit or depend upon her favors; he who asks she usually denies, and is at all times a capricious and unreliable party.

One of the difficult problems that face a man in command of a small vessel making such a voyage is the choice of a crew. He must not only have fearless, skilled men, but men who can stand discomforts and hardships without losing their tempers or giving way to despondency, and above all they must have perfect reliance on his knowledge and skill, and be willing to abide by his decisions and accept his judgments at all times. And they must keep their fears or differences to themselves, and not criticise the commander's actions or question his moves.

This gossiping and carping among crews is what has led to the fighting and failure of so many expeditions and exploring parties, and usually starts with one man. I have the same bother as I suppose every yacht skipper has with racing crews. The man who starts the trouble is invariably the most useless and ignorant of the crew. This criticism and carping is invariably born of ignorance. What at times may look to the crew of a vessel as a wrong or foolish move, may be so, if taken as a unit, but as a part of a commander's combination it may be eminently right and sane. You do not know what is in his mind, and cannot see as he sees the combination in all its phases; he is grasping and employing it as a whole, you are only looking at and comprehending a part, and that perhaps a very small one.

Nor do I think it advisable for a skipper to take his crew into his full confidence, and explain his complete plan, because it is often necessary if not imperative to make changes which cannot be foreseen, and the making of these changes always shakes a crew's belief in a skipper's skill and judgment unless they can comprehend the necessity of the altered action. I make it a practice to tell a crew only so much as is necessary to an intelligent and zealous performance of their duty, or to keep them from growing despondent and careless by anticipating no success.

A crew of three such as we had on *Sea Bird* is the safest number, because it is too small to form a clique, and too large to be unsociable. My crew on *Bird,* in the first place had implicit confidence in their commander, and the second the same confidence in the activity and skill of each other, and the skipper had a far-reaching confidence in their skill, pluck and endurance. In consequence, we three worked together like the wheels of a clock, and at no time was there jarring or rasping of metal.

They understood and were amenable to discipline, and comprehended that if things were to be run right they must be run in regular and orderly manner, and all hands making themselves subservient to the interests of the undertaking. Each man's duties were mapped out and never to my knowledge purposely neglected. Once they had their work mapped out I did not interfere, nor had I except once any cause to.

Much of the success of a venture depends upon the health and spirits of the crew, and health and spirits are more than anything else the result of regular and plenteous feeding. While it is sometimes impossible to regularly and largely feed men in land ventures, it is, except for a day or two of very bad weather, always possible at sea, as a sufficiency of stores can be carried without it imposing any burden on feeders. Men underfed or fed at irregular intervals soon become cross and dissatisfied, and this leads to despondency and ill health. Nothing so conduces to success in a venture of this kind as a regular and unstinted diet of cooked food.

While men can and will live on prepared foods, they soon grow tired of them, and it is best to have a quantity and variety of uncooked food that can be readily prepared. Two of these things we carried in abundance and at all times relished—potatoes and rice.

The consumption of fluid is another problem, and if you can carry enough to allow your men a surfeit of this very necessary food so much the better. Men do not flourish any more than plants on a shortage of water. To help out the water supply, foods containing fluid are a great help, soups in particular supplying moisture as well as nourishment. Liquors or wines should be carried, but used in sparing quantities and only when undue labor or exposure calls for a dose of alcohol. At such a time a stimulant is exceedingly valuable. The chief dependence for warmth and nourishment is to be placed on cocoa, coffee and tea.

Besides strongly and regularly feeding, great care should be taken to husband the strength of your crew as much depends on that. The general who comes to the pitched field with an army worn out with long forced marches and meets a thoroughly rested and snappy opponent is very likely to be beaten. Give your men plenty of sleeping time and see that they go below and rest. Do not weary or annoy them with unnecessary work, and if possible do not break into their watch below, especially if they are sleeping. The necessity of calling the watch below can often be guarded against by reefing or shifting sails before the relieved watch turns in. In racing this is largely impossible, for a race is a continuous battle, to be fully contested from start to finish, and nothing can be spared if you are to win, but even in such a struggle much can be done by the skipper to reserve his crew's strength and still not jeopardize his chances.

Nor should the vessel that carries you be neglected; a constant surveillance is essential to safety and success. If the weather permits, a daily inspection of sails, rigging and hull should be made, and any defects or weakness remedied at once. Do not allow your vigilance in this respect to be put to sleep by a spell of good weather, and absence of incident. Always have your arms prepared, then even the unexpected will find you ready to fight. In 32 days of sailing between the American coast and Gibraltar we never parted a rope, tore a sail or strained a spar, because the rig was kept under constant surveillance and at all times in perfect working condition.

A fallacy which is widespread and has strong hold on yachtsmen as well as landsmen, is that vessels beat to windward in making long passages at sea. Much of the spread of this nonsense is due to publishing in yachting or boating papers tommy-rot written by men who have no knowledge or experience of ocean voyaging, and accepted and printed by editors who have, if possible, less. Because men see these statements in print they accept them without question, yet if they took the trouble to think they would at once recognize their fallacy and absurdity.

Vessels beat to windward along a coast, or up a bay or through a sound because it is impossible, owing to the contracted area to do anything else, but in the open sea it would be an absurd waste of time, when by reaching or running you can make your destination sooner and much easier. A vessel at sea often stands on a wind in order to hold up into a favorable slant, and ships making the Western Ocean passage to the Westward alternately reach on opposite boards, but they do so with the wind eight points off the bow, the means of the course being a true Westerly gain.

I saw in one paper a statement that *Sea Bird* would be a long time on the passage because she would have to *tack* from the Azores to Gibraltar, and *as she was very slow sailing in the eye of the wind,* it would considerably delay her. With all due respect to the writer, allow me to say that any vessel *that sailed in the eye of the wind* would not only be delayed but would travel stern first and would never reach her destination if it lay ahead.

⚓

The Beaches of Dunkirk

'BARTIMEUS'

Neither the waters of Long Island Sound, of the Great Lakes, of Los Angeles Harbor nor of Cowes witnessed the outstanding yachting event of 1940. That honor must go to the tricky and treacherous harbor of Dunkirk, which history will mark as the scene of one of the most dramatic incidents of the "Battle of Britain."

To yachtsmen the now-famous Dunkirk Evacuation is of special significance not only as an occasion of unparalleled heroism, but as a practical demonstration of the war-time value of both the amateur seaman and the small but sturdy craft which he commands.

THE chalk cliffs of Dover made a curiously tranquil background in the early morning sunlight, with jackdaws cawing and circling, and the faint echoes of gun-fire across the Channel whispering about the escarpment.

Away to the westward, alongside the pier reaching out from what was once the Lord Warden Hotel, were the funnels and smoke of transports and a hospital ship. They were pouring ashore their khaki-clad cargoes and stretcher cases, fruits of a night's desperate garnering, and long trains crammed with men were sliding along the foreshore into the haze. But that was far away; one imagined rather than saw what was going on there. Here in the shadow of the cliffs were actualities—a destroyer limping in with a heavy list, spattered with splinter holes, making fast alongside to disembark hundreds of weary *poilus,* an armed trawler going out with a defiant toot of her siren, followed by a motor yacht painted gray. The owner-skipper, in the uniform of the Volunteer Reserve, was cleaning a revolver with a silk bandanna handkerchief. He hailed the crippled destroyer's bridge. "What's it like over there this morning?" he shouted. It was his first trip. A bandaged figure replied with an impotent movement of his hand to his ears. He was probably deaf with gunfire and bomb explosions. "Not so funny!" replied a man busy about a wire aft. "Not so bloody funny." The haze swallowed them.

The yacht-club telephone rang, and the elderly steward, unaccustomed to the sound of it, laid down his paper, removed his spectacles, and picked up the receiver. A man's voice spoke authoritatively for about a minute.

The steward said nothing. He was an old Navy man and had been a pensioner for a quarter of a century, but he recognized the note in the speaker's voice. He waited till the end of the message.

"Aye, aye, sir," he said, and then added, "There's only the one yacht here now, sir. The *Wanderer*. Motor yacht, forty feet long. There's no crew, sir. Owner's fighting in France. There's a young lady on board at this moment—"

The voice interrupted him. He listened, turning the spectacles over in his knotted fingers, staring into vacancy.

"Aye, aye, sir. I'll do what I can. Old Navy man myself. They said I was too old to fight."

There was no answer. "Hullo, sir?" Silence. He replaced the receiver.

The *Wanderer* was lying at her buoy and there was no sign of the girl. He untied the dinghy lying at the jetty and rowed alongside. At the sound of the oars as he boated them the girl's head and shoulders appeared above the companionway. She was flushed and had a scrubbing brush in her hand.

"They want her, miss," he said simply. "They rung up from the Admiralty. Proceed to Ramsgate for orders. They're taking every craft on the south coast."

She brushed a lock of hair back from her damp forehead with her forearm. "I'm single-handed," she said. "Can you run the engine if I steer?"

"You, miss?" He hadn't thought of that.

"She's full up with petrol. There's water, too, and some stuff in tins to eat. Bring some bread."

"You know what it's for, don't you, miss? They won't let a woman—"

"They needn't know," was the girl's answer. She stood motionless, thinking. The ebb tide running past the strakes of the dinghy made a little chuckling noise in the stillness.

"Bring a couple of shrapnel helmets. Get them from the A.R.P. people. . . . What about Johnnie?"

"Johnnie?" He turned that over in his mind. Johnnie was simple, but he was useful in a boat. Ashore he just sat and played with pebbles, but put him in a boat and he was all there. The club employed him to ferry people to their yachts and for attending to the moorings and odd jobs like scraping and painting. He didn't speak very plain, but after all it wasn't talk they wanted on the beaches of Dunkirk. Another aspect of the situation occurred to him. She seemed to take it for granted he was coming. "What about the club, miss? I'm the caretaker *and* steward."

She had emerged from her reverie. "The club? What the hell does the club matter?"

He grinned, showing tobacco-stained fangs. "You've said it, miss. Give me half an hour."

When he was halfway across to the jetty she hailed him again. Her clear voice was like a boy's.

"Johnnie will want a shrapnel helmet too."

He nodded; she went below and fell to mopping up the mess on the cabin floor. She had decided to give the boat a scrub-out because it occupied her mind, which, since she had had no word from France for three weeks, was inclined to imagine things. This was where they had spent the happiest hours of his leave—the happiest hours of their lives. And now, for all she knew, he was waiting on those hellish beaches, one of all those thousands of exhausted men, waiting under shell and machine-gun fire for succor from England. She flung the mop and scrubber into the bucket and jerked open a drawer. There was all his old kit: gray flannel trousers, sweater, an old shooting jacket, a yellow muffler. She would push her hair up under the shrapnel helmet. His pipes stuck in a rack over his bunk caught her eye. That would be the finishing touch. Keep one of those in her mouth when they got to Ramsgate, and talk gruff. She selected a blackened bulldog and experimented in front of the glass. It tasted utterly foul. . . .

Coming down channel, they overtook a convoy of motor yachts and followed them. She had the chart open in front of her, but the daylight was fading and there were no lights anywhere she could recognize. She had never entered Ramsgate from seaward. She listened to the drone of the engine with satisfaction. Old Ferris had been a mechanician when he served in the Navy. It wasn't so good at the start, but he was enjoying himself down in the engine room now he had picked up the hang of the thing. Every now and again he put his head out of the hatch with his spectacles on the end of his nose. "Running as sweet as a nut, miss," he announced.

"Bravo," she answered.

Johnnie sat in the bows staring at the evening star. She tried to remember why she had brought Johnnie. He worshipped her like a dog, but that wasn't the reason. It was because she felt she had no right to take an able-bodied man from his work in England; and on the spur of the moment she could think of nobody on the spot who was as handy in the boat. He and she used to take Johnnie away for the week-end sometimes. Johnnie washed up and looked after the boat when they went ashore. . . . She was one of the few people who understood what he said.

She climbed ashore in the dusk, the awful pipe clenched between her teeth, and was confronted by a man in the uniform of a lieutenant commander.

"What ship?"

"*Wanderer*." Nobody had ever called the *Wanderer* a ship before. He would have liked that.

"What is she?"

"Forty-foot motor cruiser."

"Armed?"

She shook her head. Other owners of yachts were crowding round asking for orders.

He glanced at her shrapnel helmet.

"Well, you'd better collect some rifles and life belts. First-aid outfit, too, if you haven't got it."

"Then what?" She stuck her hands in her trouser pockets, making her voice as gruff and laconic as possible.

"La Panne. Time it so as to get there in the dawn. Take off all you've got room for each trip and transfer them to something bigger. Stick it as long as you can, and good luck." He indicated a gap in the barbed wire, where she supposed rifles and life belts were obtainable, and dismissed her from his mind.

She went back to the edge of the jetty and hailed old Ferris. The harbor was crammed with the dim forms of boats maneuvring for berths alongside. Beside her on the pierhead was a soldier with a Bren gun mounted on a tripod.

"Ferris," she called down to the *Wanderer,* "come ashore with me and collect some rifles and life belts." The soldier sidled up beside her.

"Here, Skipper," he muttered, "rifles ain't no use. Take me and this Bren gun. Wait till its dark and I'll slip down and come along with you. They won't miss me till I'm back."

She grinned delightedly. He would know about rifles, too. She had never fired one in her life. "All right," she whispered. "What's your name?"

"Tanner's the name, Skipper. You're a sport." She felt a bit of a sport.

The sky line was like the edge of the Pit. To the westward the oil tanks of Dunkirk were a sullen blaze that every now and again leaped upwards like the eruption of a volcano as a shell burst in the flaming inferno. Fires glowed dully along the coast, and shore batteries blinked white flashes that reached the ear as dull reverberations like distant thunder. The searchlights wheeled about the low-lying clouds into which tracer shells were soaring.

They had solved the problem of navigating to La Panne by following a paddle steamer that had half a dozen lifeboats in tow. The whole night was full of the sound of motorboats' exhausts. There was a young moon peeping in and out of the drifting clouds, and it revealed the indistinct lines of little craft far and wide, heading in the same direction.

Johnnie sat entranced by the spectacle, crowing huskily at intervals. Tanner, having mounted his Bren gun in the stern, gave her a relief at the wheel. He said it was much the same as driving a car. She practised loading the rifle under his tuition. Old Ferris visited them at intervals, calling her "Skipper." It didn't matter what Johnnie called her, because nobody could understand what he said.

"You're a bit young for this game, eh, Skipper?" asked Tanner. "How old are you?"

"About a hundred," she replied with a gruff laugh. And in that moment, before the dawn of hell's delight, she felt it.

The dawn came slowly, revealing the small craft of the south coast of England covering the Channel like water beetles on the surface of a pool. Pleasure steamers and yachts, barges, scoots, wherries, lifeboats, motorboats, rowing boats and canoes. . . . Fishermen, yachtsmen, long-shoremen, men who had never been afloat in their lives, millionaires and the very poor, elderly men and lads in their teens, answering in a headlong rush the appeal for boats. Boats for the beaches and the last of the Expeditionary Force.

Somehow she hadn't thought about the dead. Her thoughts were entirely occupied with the living. It wasn't till Johnnie began making queer noises of distress and pointing down into the shallow water that she saw them—the men who had been machine-gunned in the shallows, wading out into the water to reach security. They were still there, some floating, some submerged; in an odd way they seemed to convey resentment at the disturbance of their oblivion by the passing keels.

She called Johnnie to her side. "Take the lead line and sound over the bows. Call the soundings. Nothing else matters. Do you understand, Johnnie? Nothing else matters. I am here."

He made guttural noises, pointing at Tanner, who was blazing away with the Bren gun at a Heinkel overhead that had bombed a trawler astern of them. She held him with her eyes. "Nothing else matters, do you understand?" He picked up the lead line and went forward obediently. She put her lips to the voice pipe. "Go very slow, Ferris."

"Go very slow," repeated the old man.

She crept inshore. The beach was pitted with shell craters out of which men came running, wading out into the water to meet them. From the sand dunes more men stumbled, helping the wounded. The whole foreshore was alive with men and boats, and the smoke from the Dunkirk fires flowed over them like a dark river.

At three and a half feet she would stop. It was the least they could float in. She listened to the strange cries Johnnie emitted as he hauled in the dripping lead line, understanding them perfectly.

Presently, her mouth to the voice pipe, she gave the order to stop. Tanner was having trouble with the Bren gun and swearing in a ceaseless flow of incomprehensible blasphemy. Old Ferris, complete in shrapnel helmet and life belt, climbed out of his hatch and came towards her, lighting his pipe.

"They said I was too old to fight, but—"

"Get back. We're in four feet. I must keep working the engines." A bomb burst among the men wading towards them. She shut her eyes for a moment. "Keep on sounding, Johnnie. What water have you got?"

"Fraghfaph-ah-ah," crowed Johnnie.

"Good boy. Keep it going."

The Bren gun broke out afresh. Tanner, having cleared the jam, opened fire again, chanting oaths like a denunciatory psalm. "Slow astern, Ferris."

Another cluster of men wading to their armpits had reached them.

Johnnie looked back at her and pointed at their sun-scorched, puffing faces. No doubt existed in his mind that it was all something to do with his lead-line achievements. He was delighted. Somewhere out of sight a German field-gun battery opened fire, the shells whistling viciously overhead.

She searched every face as they came splashing and gasping towards her and somehow contrived to hoist each other inboard. She took sixty or seventy at a trip and transferred them to the nearest vessel lying out in the deep water; she had hitherto believed that the utmost capacity of the *Wanderer* was a dozen. Backwards and forwards they went under exploding bombs, under machine-gun fire and whining shells. Tanner ran out of ammunition and they went alongside a destroyer, where he got another case and a spare barrel for the Bren gun. She lost all count of time, all fear, all feeling. Sometimes she interrogated weary men: Had they seen his unit? Had they ever heard his name? They shook their heads and begged for water. She had none left.

Then suddenly it seemed that the beaches were empty. She didn't know that the men were being marched westward to Dunkirk, where the French and British destroyers were crowding alongside the mole and embarking troops in thousands under shellfire. Except for a few scattered units moving west, the beaches were empty. The task was done; but where was he— Oh Christ, where was he?

The Bren gun had been silent for a long time, but she hadn't noticed. Now, turning to look seaward, she saw Tanner lying beside it with his knees screwed up into his belly. She ran aft and knelt beside him.

His eyes sought hers out of his gray face. "I bought it, Skipper. Sorry. . . . Got a drop of water?"

She raised his head and held it against her breast. "There isn't any water left."

His eyes were suddenly puzzled. . . . He moved his head sideways a little and then smiled, and died, ineffably content.

They followed a big gray coaster back to Dover. Old Ferris got a spare red ensign out of the locker and tucked Tanner up in it. He didn't mind Tanner's being killed, having been disposed to regard him jealously as an intruder into a nice little family party. Moreover, he disapproved of his language. He walked forward to the wheel-house. She was moving the spokes of the wheel slowly between her blistered hands. Her shrapnel helmet lay on the chart beside the valiant briar pipe. She was aware of the old man beside her and of having reached the end of her tether at one and the same moment.

Old Ferris kicked Johnnie, asleep at her feet, into wakefulness. "Take the wheel," he said gruffly, and held her as she pitched, sobbing and exhausted, into his arms.

They berthed alongside the Admiralty pier and she climbed ashore to find someone who could give them fuel and water. The quays were thronged with troops in thousands, being fed and sorted out into units and entrained. A hospital ship was evacuating wounded into fleets of ambulances. She stepped aside to give room to the bearers of a stretcher and glanced at the face on the pillow.

He had a bandage round his head and opened his eyes suddenly on her face.

"I've been looking for you," she announced in a calm matter-of-fact tone. She felt no emotion whatever.

He smiled. "Well, here I am," he said.

⚓

On the History of Knots and Rope Making

RAOUL GRAUMONT AND JOHN HENSEL

Down through the ages the tying of knots and the making of rope have played highly important roles in the life of man. That there has always been a need for rope or cord of some kind can be readily appreciated. Since rope could have served but few useful purposes unless it could have attached in some manner to the things it was desired to pull, lift, or secure, man, at the time of his first conception of the use of rope, must have conceived of some means of tying knots.

History tells us that the first cords were made from the tendrils of vines, from the cord-like fibers of plants, and from strips cut from the skins of animals. When the vine tendrils and the cordlike fibers which were readily available failed to serve their immediate needs the primitive peoples of the world began to weave, twist, or braid these strands of fibers to make ropes of greater strength and added length.

Records are available to indicate that most of the ancient civilized nations of the world were accomplished rope makers, as were many of the savage tribes of the globe. Each of these races and tribes made use of the materials which were most easily accessible. These have been found to include the fibers of many different kinds of plants, the skins and sinews of animals, and the hair of both humans and animals.

Specimens of rope made by the early Egyptians of flax, papyrus, and rawhide have been found in tombs estimated to be not less than 3500 years old, while it is generally well known that rope was made in China at a very remote period.

Neolithic man made rope and tied simple knots. The lake-dwellers during the Stone Age, and the Incas in Peru used the sheet-bend in making their nets. The Incas also had a decimal system of numbers based on knots tied in suspended cords, the type of the knot and its position in the cord each having a special significance.

Among some of the most interesting relics of these ancient Incas are these so-called *quipus,* or knot-records. It appears that this race of people never discovered the art of writing but that they developed a decimal system of numbers and with the aid of knots tied in cords were able to keep records of large sums and figures. It is supposed that they may have used their system of knots for difficult mathematical calculations, but this has not been determined definitely. Their method was substantially as follows: Several vertical cords were suspended from a horizontal cord and in each of the vertical cords knots were tied, the whole being called a *quipu.* It is assumed that a single overhand knot represented the figure one, a double overhand knot, two, and so on, up to nine. It has been found, too, that the figure-of-eight knot and the simple running knot were used and had particular significance. The knots representing the units were tied in the lower ends of the pendant cords, with the tens immediately above the units, the hundreds above the tens and the thousands above the hundreds. In this manner dates, astronomical records and other large numbers could be preserved.

In the historical records of the early Greeks and Romans are to be found many examples of the use of rope. As another example of early rope making a painting of a drinking cup in the British Museum shows an Attic sailing ship of about the sixth century fitted with sails and ropes. The Indians of North America were also accomplished rope makers and used knots for recording dates, while from the sailor's point of view some very interesting specimens of aboriginal skill come from the Nootka and Clayoquot Indian tribes of Vancouver Island and the coastal regions of the state of Washington. These peoples made whaling lines out of small cedar limbs twisted into three-strand rope which was about four or five inches in circumference. One of these lines is known to have been 1200 feet in length. For the lanyards of their whale harpoons they employed the sinews of the whale, twisted into three-strand rope.

It is not definitely known whether the art of making rope was transmitted through the channels of trade and social intercourse, from one race of people to another, or whether it was independently evolved by the various peoples of the globe as they emerged from savagery. All that can be stated about this matter is that rope and cordage of some kind was one of man's earliest and most useful tools, that its use was widespread, and that a high degree of skill in the making of rope was achieved long before the dawn of history.

That the reef knot was well known in ancient Greece and Rome may be seen in many existing works of the classical art. It was used in an ornamental manner in the handles of vessels, while it nearly always appeared on the staff of Mercury and in many instances in pieces of sculpture in the girdles of Roman vestals. These people called it the knot of Hercules (that is, *nodus Herculis* or *Herculaneus*), because they thought it had been originally tied by Hercules. One historian declares that it was the custom for Roman brides to wear a girdle tied with a Hercules knot, which their husbands untied on the marriage night, as an omen of fecundity. Pliny states in his *Natural History* that it was found that wounds healed much more rapidly when the bandages which bound them were tied with Hercules knots.

In some writings by Oribasius there are descriptions of some eighteen different kinds of knots used in surgery, but unfortunately the Greek names he employs do not serve to identify the knots at this time. Probably one of the earliest knots familiar to the English people was the Carrick bend. This was used as an heraldic badge by Hereward Wake, the Saxon leader who refused to submit to William the Conqueror in 1066 A.D. This knot is now commonly called the Wake knot.

The names of the knots as we know them today afford no clues as to their origin. The familiar reef knot was known to man long before the origin of the English word reef. We are told that the word knot comes from the Old English *cnotta,* meaning to join together, while the sailors' "bend" comes from the Anglo-Saxon word *bygan,* which is to make a loop or fashion a "bight" in a rope. There are lovers' knots, matrimonial knots, the Gordian knot and others whose names are but figurative. The Stafford knot, the Bouchier knot, the Henage knot and others are in no sense knots, but badges of heraldry. Then, too, there is the monkey knot, the Peruvian knot and the wind knot, all of which are fancy knots.

Among the real and useful knots are the weavers' knots, the builders' knots, the sailors' knots and the fishermen's knots. Included in the classification with knots are bends, hitches, lashings, seizings, whippings, shortenings, stopperings, bowlines and splicing, which find many uses in everyday life.

The various methods by which any of these knots are tied are in large measure dependent upon three primary conditions; the material available, i.e., the rope or cord to be used; the purpose for which the knot is intended, and the ability to so manipulate the rope or cord into the form of knot desired.

In his famous *Sea Grammar,* published by Captain John Smith about 1627, he asserts that the three commonest knots used by sailors of that time were the "Boling" knot, the "Shepshanke" and the wall knot, although he does not make mention of any of the bends, hitches and splices which surely must have been known at the time his *Grammar* was published.

There are countless stories told, purporting to relate the origin of many of our commonest knots. One of these is offered as the source of the name of the thief knot. It has to do with a tale told about an old Cape Cod sea captain who suspected one member of his crew of stealing bread from the captain's breadbag. One night, instead of tying the bag with a reef knot, as had been the usual custom of the captain, he tied it with a thief knot instead. Since the two knots are very similar the wily captain anticipated that the thief would tie the bag again with a reef knot (which he did). In this manner the thief was found out, and, so it is said, since that time the knot has been known by its present name. As a matter of fact, however, the thief knot was known by many different peoples in many different lands long before there were any Cape Cod sea captains.

Among many of the primitive peoples of the world knots have often been associated with magical and supernatural powers. Wizards and witches, or persons who were regarded as such, have been known to contend that they had the power to tie up the wind in knots. And mariners, who as a class are notoriously superstitious, bought these charmed cords of knots to be untied when they were becalmed at sea. There were supposedly three of these knots in the cord, which when untied in order had the property to release a wind of moderate force, then a half-gale and finally a blast of hurricane proportions. Such superstitious beliefs as these were prevalent in Lapland, Finland, Shetland, the Isle of Man, and other northern countries, particularly among seafaring men. That the same superstitious beliefs were held by the early Greeks is evidenced in Homer's epic in which Ulysses was presented by Aeolus, king of the winds, with all of the winds tied up in a leather bag.

And so, throughout all history, and among the records of many civilized nations and savage tribes as well in all sections of the world there are many references to ropes and knots.

⚓

~~~~~~~~~~~~~~~~~~ *45* ~~~~~~~~~~~~~~~

Northward in "Pinta"

DENNIS PULESTON

PINTA was a 57-foot gaff-rigged schooner, designed by John Alden. She had a great reputation as an ocean racer, for she had finished a very close second in the famous King of Spain's Transatlantic Cup Race of 1928.

Often, when reviewing the characteristics of small sailing ships, I have tried to find their counterparts in the animal world. *Pinta,* graceful and speedy, reminded me of a thoroughbred race horse. She was a creature of moods, and at times required very careful handling. She could not be left to her own devices, but must be watched and tended constantly. Yet if all her caprices were humored, she was a safe, fast, but not too comfortable creature. She was very well equipped and, compared with the close quarters on *Uldra,* a veritable palace below decks. Electric lights, a toilet and a washbasin were luxuries we had not known. Forward of the main cabin was a neat galley, with a gas cooking range, an oven and a sink.

When we entered St. Thomas Harbor, we found anchored there a magnificent sailing ship. She was the *Suomen Joutsen,* or White Swan of Finland, a full-rigged ship manned by cadets of the Finnish navy. There are few square-riggers now left, and their owners cannot afford to maintain them in the splendid style of the times when sail was in its heyday. But this one was nothing less than a naval vessel and was kept in a style worthy of her position. She was manned by a crew of two hundred and fifty odd, who kept her looking like some glorious great yacht. Her canvas was spotless, her spars gleamed with varnish, her

Reprinted from *Blue Water Vagabond* by permission of, and copyright 1939, by Double-day, Doran & Company, Inc.

brasswork winked at us in the glaring sunshine, and her sides were white as any swan's.

No sooner were we anchored off the town than a party of her officers came over and invited us aboard. They showed us everything, from her spacious cold-storage holds to her luxurious saloon. We were still on board her when a strange feeling came over me. I started to shiver, the whole scene swam around me, and I was seized with a nausea. I hurried back to *Pinta,* crawled into my bunk and within ten minutes was running with sweat, in the first stages of malaria. It was ironical that I had escaped it for many months, through the worst fever season of the year, only to fall a victim just as I was on the point of getting away to sea, beyond its reach.

For several days I sweated and shivered. When we finally sailed, the fever had left me, but I was too weak to come on deck. I saw nothing of the run down to Puerto Rico. For once I was a passenger, though a very unwilling one. I dozed, and when I awoke, it was another day, and we were in the harbor of San Juan, snugly moored alongside the dock. We stayed there several days, and by the time we left I was able to lend a somewhat feeble hand on deck.

We sailed from Haiti for the Bahamas in an exceptionally strong breeze. Pinta seemed to enjoy the change from the usual mild Trade wind. Outside the harbor we bore away to pass to the westward of Great Inagua Island, and with the wind just abaft the beam she tore along like a frightened stag. She was far from easy to steer and was taking over sheets of spray, but in the thrill of driving the little schooner to her utmost, we did not mind a few discomforts. We passed to windward of Tortuga Island, one-time stronghold of the pirate Morgan, and it disappeared in a howling black squall. Our run for the past twenty-four hours was 223 miles, an average of well over nine knots. Great little ship! That was a sail I shall remember all my life.

For the rest of the passage we had light breezes, and cruised leisurely through the Bahamas. They are a strange group of islands. None of them is more than a few hundred feet above sea level; most of them are mere flat slabs of limestone rock that have been pushed up out of the sea and are now covered with low scrub. Between many of them are great stretches of shallow water, with a bottom of clear white coral sand. Some of these shoals, locally known as "Banks," cover hundreds of square miles of water. On them the water is a brilliant emerald green, save for the occasional brown patches of coral which have grown up like trees. So dazzling are the colors of the Banks that when the sun is shining on them they are reflected in the clouds overhead. Far off at sea, before the islands have been sighted, the strange green glow of the Banks can be seen in the skies.

One fine moonlit night we reached Nassau, chief town of the Bahamas, and anchored in the fairway. It is a smart and prosperous place with a busy harbor. Sponge-fishing and trading schooners and cutters, manned by crews of jolly, noisy Negroes, were constantly arriving and leaving for the out-islands. In Nassau I had my introduction to the diving helmet. Johnny had one aboard, which so far we had not used. But we were determined to see the far-famed coral gardens of the Bahamas. Rather like the casque of a medieval knight, the diving helmet fitted loosely over the head and rested on the shoulders. It was made of copper, with lead weights of about sixty pounds attached, and had a glass window in front. Air was pumped into it through a rubber hose from a small hand pump from on deck. The surplus air escaped from underneath the helmet around the shoulders, in a steady stream of bubbles. We found that in greater depths than twenty feet it was much easier to walk on the bottom with heavy boots, since the feet have a tendency to rise.

This was opening up a new vista in my seagoing life. After my first descent I wished I could spend all day in exploring that strange new world. I wanted to know more about the little fish, brighter than hummingbirds and quite as fearless, that people the coral trees in thousands. I saw the dangerous moray eels that lurk in the darkest caverns; the huge and hideous brown groupers, with their slow, deliberate movements; and the lean, steely-gray barracoutas, hawks of the tropical seas, who strike at their prey with the speed of a whiplash. Once I saw a large octopus, his greenish skin veined like marble, an evil spirit squatting in the mouth of his coral cave. He stared balefully at me with his cold eyes, and even as I watched him, his colors changed. With magical suddenness he had turned to a pale brown, with faint red blotches; then to a uniform dirty gray; and then to a light yellow, with umber patches. I prodded at him with the butt end of the fish spear, and in a moment he had ejected a dense cloud of sepia. When it cleared, he had vanished. Down in five fathoms of water we found the bones of an old ship, her ribs sticking out of the soft sand. Once she had been a proud schooner; a "down-Easter," our Bahaman launchman told us, in the Caribbean logwood trade, until a hurricane had claimed her. Now she was a playground for the schools of little black-and-yellow striped zebra fish.

But we had work to do down there. The conditions were ideal for submarine photography, and after we had all made ourselves familiar with the diving gear, we started our experiments. Someone had hit on the brilliant idea of using a pressure cooker as the container for the underwater camera. These are strongly made, built to hold a heavy pressure of steam, so we were sure that no water could leak in. We had fitted a glass aperture in one end of it for the lens and had an electrical attachment to operate the starter. With chopped-up conch meat we attracted schools of fish in front of the camera and were able to obtain many good pictures of them in their natural colors.

Diving with simple gear in tropical waters is becoming an increasingly popular form of sport. Dr. William Beebe, the great authority on oceanographic research, has done much to popularize it. A helmet, a pump, a length of hose, a bathing costume and a pair of rubber shoes —that is all the equipment needed to open up a new and intensely beautiful world. I consider that diving gear should be included in the equipment of every yacht cruising in Trade wind latitudes. And not only will it prove a source of interest and pleasure on many occasions, but it will be of use in scrubbing weed from the bottom of the ship, or in effecting minor underwater repairs to rudder or propeller, in seas where it is often many hundreds of miles to the nearest drydock.

Before we sailed from Nassau, there were two additions to *Pinta's* crew. Coulton Waugh and Ellsworth Ford had taken a holiday and had come down from New York to help sail the ship home. They are both yachtsmen of distinction. Everyone who has sailed to any extent on the east coast of America will know them. "Lank" Ford, with his inseparable accordion, is to be found on every ocean race. His tall figure, which has earned him his nickname, is usually adorned with all manner of remarkable garments; in *Pinta* he wore brick-red Basque pants, a Breton fisherman's smock and a scarlet fez. On his face is an expression calm and benevolent; during a long friendship with Lank, I have never seen him flustered. He is always cool, and in times of emergency on board ship he has a happy knack of being just where he is most needed. It is small wonder that he is in great demand on the ocean racers; for not only is he one of the finest amateur racing hands, but he can play the accordion like a maestro. Coulton, artist, writer, seaman, and musician also, was another extremely useful hand.

So, when *Pinta* got under way for America one bright morning, she was manned by as useful a crew as could be found anywhere. We were looking forward to a fast and easy passage. There were no premonitions of coming disaster in our minds. Although after the tragedy Emil, our cook and paid hand, told us he had been having evil dreams for some time, we told him somewhat unsympathetically that they were caused by overeating. And then he pointed out several other omens. It would never have happened, he said darkly, if only we had refrained from whistling before breakfast. That is a fatal thing to do at sea.

We were a gay crew that day as we sailed out past the Hog Island light and headed north to pass to windward of Abaco. There were Johnny and Emil, excited at the thought of returning to their homes after many months of cruising in the islands; there were Lank and Coulton, looking forward to a sail in tropical waters after the unpleasant Northern winter; and there were Geoff Owen and I, eager to see a great new country. Even Tiger, our dog, frisking about after a rope's end Emil was dangling in front of her, seemed more kittenish than usual.

As soon as we were well out to sea, on our course, and had all halyards coiled down, Lank brought out his accordion and played us a lively tune. We watched the neat little town of Nassau sink down below the horizon, and by dusk had picked up the blink of the lighthouse at Hole-in-the-Wall, on Abaco Island. Luck seemed to be with us. The belt of calms and variables which often lurks north of the Bahamas was not in evidence. For the next three days we scudded northward with a fresh east wind and were soon driving through the cobalt waters of the Gulf Stream, in a series of stiff squalls.

Next morning, with startling suddenness, we ran out of the Gulf Stream. One moment, stripped to the skin, we were reveling in the warm air, when all at once the color of the sea changed to a cold green. We began to shiver and dived below for our sweaters and long pants. Within the space of a minute we had sailed out of the tropics into early northern springtime.

Just about this time we noticed a bird sitting unconcernedly on the forward hatch. He was evidently very weary, for he did not attempt to fly away when Lank approached him. He was a strange and gaudy-looking creature, a purple gallinule, as we afterward learned. But none of us knew what he was then, and to Emil he was an omen of evil things to come. We put him in a basket where he could regain his strength, safe from the flying spray, and for the time forgot about him.

In the early afternoon we sighted the Diamond Shoal lightship dead ahead—a perfect landfall. We carried on toward the North Carolina shore and just before dusk could make it out, a low line of dunes, surmounted by the tall, barred spike of Cape Hatteras lighthouse. It had been decided that we put into Norfolk, at the mouth of Chesapeake Bay; so we carried on up the coast, laying our course parallel with it, and about eight miles off shore. It looked as if it would be a fine night, with a moderate easterly wind, but soon after dusk the sky clouded over, and it began to drizzle.

Soon after ten o'clock that night, out of the blackness to seaward, a vicious squall rushed down on us. One moment the ship was reaching quietly along, the next she was plunging madly before a screaming wind. Her lee decks were buried under a hissing smother of phosphorescent foam; sheets of spray rose from under her weather bow and went hurtling aft the length of the ship. It was time to take in sail and to put the ship about, so the watch on deck sent a man below to muster all hands.

Just as we were coming up through the hatch, she struck. She shuddered violently, then lay over almost on her beam ends. I remember my first impression: that a great sea had hit her, and she was going to roll right over. But when I scrambled up on deck, I realized at once that she was aground. No longer was she surging ahead through the water. Just then she came down with a crash on something hard. From the galley came the muffled clatter of falling pots and pans.

"We're on the beach!" said Johnny Green as he grabbed the wheel and twirled the spokes in a desperate effort to get the ship around. Emil darted down into the engine room and started up the engine. I let fly the headsail sheets, the others began lowering the foresail; in a moment everyone was at his post, instinctively and without panic doing what he could to save the ship. But we soon saw our efforts would be of no avail. Each sea which struck us was driving us farther inshore;

each time we bumped, it was with greater violence. It seemed as though it would only be a matter of minutes before she broke up unless—we hardly dared hope—unless she bumped right over the shoal on which she lay, into deep water on the other side. But we were sure she could not stand much of this battering; it was a most horrible sensation to feel that little ship quivering pitifully under each blow, while we were powerless to help her. We could feel her timbers crack and groan in agony.

In the meantime the squall was at its height. Great black seas were rearing up at us out of the darkness, to strike against the side of the ship and hurl their spray savagely at us. Before we could muffle the headsails, they had beaten themselves to tatters, and the fierce wind was still worrying the last rags of them. Somehow we had managed to lower the mainsail and foresail; now all we could do was to wait and wonder what would be the fate of our ship and of us.

None of us could believe that we had been carried so far off our course as to be aground on the coast of the mainland. Therefore, we reasoned, we must be on some off-lying shoal, although we did not know of one marked on the chart. There was nothing in that dismal darkness to indicate where we lay. And what if we were wrecked many miles from shore? We would have to make a dangerous trip through the heavy sea, in our dory, or perhaps be drowned.

Lank went below and coolly produced the Very pistol. Up into the blackness went a red flare; then, several moments later, another one. As this second one came down, our hearts leapt. For we saw, lit up by its ghostly glow, a shining bank of sand. At least there was land of a sort over there. Just then *Pinta* trembled for the last time, and her struggles ceased. Driven shorewards by the wind and sea, she was now hard and fast aground, and her sufferings from the cruel battering of the breakers were over for the time, at least. We could only guess at the extent of her damages, but it seemed highly probable that at the next high tide she would break up. We must try to save what we could. Everyone started gathering together his most important possessions.

My first thought was for Tiger. By this time she was thoroughly frightened. Her well-developed sea sense told her that something was very wrong. The shots from the Very pistol had further distressed her gentle soul. I picked her up from where she cowered trembling in the corner of my bunk and carried her on deck. I climbed down the steeply sloping deck to the lee shrouds and went overside into the chilly water, which only came up to my middle. Just as I was halfway in to the shore, a big breaker curled up over my head, knocked me down and turned me over. Poor Tiger slipped from my arms; when I rose spluttering to my feet and looked for her, she was nowhere to be seen. Then I heard a faint whine from behind me. Tiger had swum back to her ship and was trying to scramble aboard. So I had to begin my rescue act all over again. This time we reached shore safely. I set her down; she shook herself and then ran off into the night.

I found myself on a beach of sand, wet and firm from the sea. But after I had taken a few steps, it became dry and soft with here and there a tuft of wiry grass.

A Coast Guard patrol from the nearest station, only two miles away, saw the flares and reported them to his mates immediately. In a very short time a truck came tearing along the beach, and half-a-dozen men were helping us salvage our gear. They summed up the situation at once.

"I reckon we can't do anything for the ship until daylight," said one. "Maybe she's hurt, and maybe she ain't. She's set hard in the sand now, and we'll see when the tide is in again. In case she is, we'll take off all your stuff now."

They also explained how we had come to be driven so far off our course. The currents close to Cape Hatteras are strong and uncertain. On some occasions, they said, there is a strong current which sets directly on shore, and which may run as hard as three knots. That is what had tricked us.

The morning dawned fine and mild. The Coast Guard, as always, were up with the sun. We drove along to poor *Pinta*. She was a sorry sight as she lay there inert, far over on her side, with the rags of her headsails trailing in the sea like a dead woman's hair. She did not look as if she would ever again be the lively, eager little ship we knew. But after a long examination the Coast Guards surprised us.

"She seems all right," said Captain John Midgette, the "boss" of the station, "and I think we can get her off tonight."

"But how?" we began.

"There's our big cutter, the *Mascoutin,* up at Norfolk," replied the captain. "I've already phoned for her, and she'll be along this afternoon. Then, when the tide is high tonight, we'll try and pull her off."

On our way back to the station we felt a great deal better about the whole affair. We realized that we were far from being the only ones who have blundered onto that dangerous coast. The captain pointed out to us the spots where, in his memory, ships had come ashore. In some cases there was still a battered hulk, a mast sticking out of the sea or a few weathered ribs half buried in the sand. But often there is nothing left, only a memory. From below Hatteras, right up to the Virginia Capes, off the mouth of the Chesapeake, he told us there is a wreck for every hundred yards of coast-line. "The Graveyard of Ships" it has been aptly named. Beset by uncertain currents, sudden squalls, violent storms and fogs, it is merely a chain of low-lying sand dunes, in many places not more than a quarter of a mile wide. On it the Coast Guards and their families lead a hard and lonely life. Before them the Atlantic hammers unceasingly at their feet.

After lunch a smudge of smoke was reported away to the northward. "Here comes the *Mascoutin*," said Captain John. "We'll go along and meet her now." So we returned to *Pinta*. This time our truck towed a large trailer, on which rested the station's sturdy surfboat. When the *Mascoutin* reached a point about half a mile off the coast where the wreck lay, she anchored. The Coast Guards ran the trailer down to the water's edge, and the surfboat was launched. There was a long stretch of breakers through which they had to go, and quite often they disappeared from sight behind a great wall of surf, but Captain John steered her with expert touch, the men rowed steadily and untiringly, and the little bobbing craft did not ship more than a few cupfuls of spray. Soon they were beyond the broken water and had reached the side of the *Mascoutin*. She passed them the end of a light line, which they brought back ashore with them.

Night fell, and still we waited tensely for the tide. The truck was turned around, and her headlights were focused on *Pinta*. We watched the breakers gradually reach higher and higher about her hull. Suddenly one of the Coast Guards gave a grunt. She had moved! Ever so slightly she had stirred, as if awakening from a deep sleep. Another twenty minutes dragged by. She was now rocking regularly as each sea swirled about her. Then from the *Mascoutin* came a series of flashes. The time had come!

We saw the cable gradually tauten and stretch. It did not seem possible that any rope could take such a strain. But *Pinta* started to turn, until her stern was facing directly out to sea. All at once she straightened up on an even keel and began to move slowly seaward in a series of bumps. We were breathless with excitement; every inch she moved now, she was getting into deeper water, and soon her keel would be free of the sand.

Suddenly, above the steady roar of the surf, we heard a sharp crack. Something struck the sea beside *Pinta* with a splash, and she lay once more on her side. Emil groaned.

"It's no use, boys! The cable's broke!" he said with an air of finality.

Our hearts sank. The Coast Guards had done their best to save our ship, but had failed. I remembered the bleached bones of the half-forgotten ships I had seen that morning, and thought of *Pinta* sharing their graveyard. Soon she, too, would be just a name and a few timbers sticking out of the sand. But what were the Coast Guards doing in the meantime? While we stood dismayed at the catastrophe, they sprang into action. There was a sharp order from Captain John Midgette, and the surfboat was run down and launched once again. They were not beaten yet. In the eerie light which the truck was throwing out over the water, we watched them dip and rise again among the breakers. It was a sight such as few have been privileged to see. Our spirits rose once more, for it did not seem possible that such courage and perseverance could fail to win in the end.

Soon they had disappeared into the darkness beyond the range of the lamps, and we could only wait. At length: "Here they come!" exclaimed Johnny, and we saw their dim shape again. They were bringing back the end of the hawser with them. Somehow, in the midst of the surf, they joined it with the shoreward end. It was a piece of small-boat seamanship I do not think I shall ever see equaled. A flash from their signal lamp at last, and once more the *Mascoutin* took the strain. This time there was no hesitation. *Pinta* rose, shuddered three times as she struck bottom in the troughs, and was free. Soon she had vanished into the night. There came a flashed signal from the *Mascoutin*.

"She says the little schooner is doing all right," translated Captain John. "They'll put a couple of men aboard her to keep her pump going, and will tow her up to Norfolk. There's nothing more we can do now. In the morning you fellers can go on up there and see how she is."

We returned to the station, our hearts too full just then to express our thanks and admiration to the captain and his men.

Next morning we left, and by ferryboat and bus reached Norfolk. We could hardly believe the report of the shipwright who had surveyed the damage. He told us that she had suffered little.

Pinta has had a strange life. Once she was squeezed severely between two steamers. Once she caught fire. Many times has she been aground, her hull bumped and battered against rock and sand. But she has survived it all. On Long Island Sound yachtsmen call her "the rubber ship." Today she is as trim and sleek as ever, and still sails with a jaunty air, which no disaster has been able to take from her.

⚓

From a Navigator's Note Book

ANTHONY ANABLE

DISCOVERING LONGITUDE

STRANGE as it may seem, reliable methods of determining the longitude of a ship at sea are of comparatively recent origin. This would appear to give credence to the quip often heard that "When Columbus set out he didn't know where he was going; when he got there he didn't know where he was; when he got home he didn't know where he had been." The same probably goes for Magellan, Cabot and all the other early navigators.

They all knew how to determine latitude by the ancient Meridian Altitude of the sun at high noon—the oldest and simplest element of both the old and the new navigation. And like many a shipmaster of comparatively recent times, they could make port with a dead reckoning guess at their longitude by the simple expedient of getting on the latitude of the landfall they sought and running due east or west until their destination loomed up on the horizon.

It wasn't until 1773 that the inventor of a reliable method of determining longitude received the £20,000 prize posted in 1714 in England by a learned body called the "Commissioners for the Discovery of Longitude at Sea." The magnitude of the prize award is evidence of the importance placed on the solution by the greatest sea power in the world and the general dissatisfaction with the cumbersome and impractical method then in limited use, called the method of Lunar Distances, which is based on the changing position of the moon during the fixed stars.

The test to determine the winner of this munificent award was an essentially practical one and one designed to deter the crackpot and long-haired theorist. The decision was to rest upon the contestant's ability to direct a vessel home to a designated port after having been

Originally published in and copyrighted by *The Sea Chest: The Yachtsman's Digest.*

shipped in her blindfolded from an unknown port and kept confined to a stateroom for whatever period of days was required to make a good offing on a secret and intricate series of courses. It was a minutely regulated game of blind man's buff with a £20,000 bounty for the winner, at the discretion of the venerable commissioners.

Betting ran high on the probable winner and Maskelyne, the Astronomer Royal, was an odds on favorite. Maskelyne based his hopes on refinements in the cumbersome Lunar Distance method. But to the surprise of the "Commissioners for the Discovery of Longitude at Sea" the solution was obtained by a little ticking thing in a box, the work of a Yorkshire carpenter, John Harrison, to whose son, William, after much quibbling, delay and procrastination, the award was finally completed in 1773, after the death of the inventor. Thus the chronometer, the first accurate time piece, was created and longitude was "discovered."

Harrison's solution was based on the fact, known even before the day of Columbus, that longitude at sea was determinable if one knew the exact time at some standard meridian such as that of Greenwich, England. Up to Harrison's time no accurate time piece existed, with which to carry Greenwich time to sea.

Longitude east or west of the meridian of Greenwich is readily converted into difference in time between Greenwich, England, and the position of the ship, for 15 degrees of longitude equals one hour's difference in time. A Time Sight, taken preferably on the sun when it is bearing true east or true west from the ship gives the time at the ship. So, if the shipmaster carries along with him Greenwich time ticking merrily away in a little wooden box, the longitude is his by simply taking the difference between Greenwich time, shown by his chronometer, and ship's time as computed from his Time Sight. Simple, assuredly. So simple, in fact, that it took a carpenter to figure it out while an Astronomer Royal and a galaxy of mathematicians went about it the hard way.

THE DISCOVERY OF "LINE OF POSITION" NAVIGATION

The pages of maritime history contain few more dramatic incidents than Captain Thomas H. Sumner's simple account of how he discovered and applied the Line of Position Method on December 18th, 1837, off the east coast of Ireland. To the navigator, amateur as well as professional, the incident is epoch making in its importance, truly the birth of the "new navigation." Even to the laymen the drama inherent in the situation is apparent and is heightened rather than lessened by the simple, seaman-like manner in which Captain Sumner logs the event.

Up to this time navigators had relied solely on Time Sights for longitude, after the invention of the chronometer by John Harrison in the previous century. But the Time Sight method used by Sumner on that day was dependent on knowing latitude exactly and consequently was only to be relied upon as an approximation when dead reckoning latitude was in doubt, as frequently was the case.

The Sumner Line, forerunner of the Marc St. Hilaire and twenty-nine other line of position methods, conferred the greatest boon on navigators since Nathaniel Bowditch published in 1802 his first volume of navigational tables. Thanks to Sumner and those who followed him, a navigator today may take a sight at any convenient time and establish his position somewhere on a line drawn on his chart, regardless of any reasonable error in his dead reckoning position. One sight yields one line of position, two sights two, three sights three. The intersection of the lines obviously determines the "fix" or true position of the vessel.

Captain Sumner was not a mathematician like Bowditch. He was simply an intelligent, practical American shipmaster on a more or less routine voyage from Charleston, S. C., to Greenock, Scotland, via the Irish Sea. For several days he had been cursed with poor visibility and a complete lack of sights. Uncertainty as to his position and the approach of a storm instilled in this prudent seaman a well reasoned caution as he entered the Saint George channel between Ireland and Wales, and closed with the east coast of Ireland to leeward, near Tuskar Rock. So let him tell his story in his own words, just as he jotted it down in his log book and as it appears in *The American Practical Navigator,* the current edition of "Bowditch" published by the Hydrographic Office.

"Having sailed from Charleston, S. C., 25th November, 1837, bound for Greenock, a series of heavy gales from the westward promised a quick passage; after passing the Azores the wind prevailed from the southward, with thick weather; after passing longitude 21° W. no observation was had until near the land, but soundings were had not far, as was supposed, from the bank.

The weather was now more boisterous, and very thick, and the wind still southerly; arriving about midnight, 17th December, within 40 miles, by dead reckoning, of Tuskar light, the wind hauled SE. true, making the Irish coast a lee shore; the ship was then kept close to the wind and several tacks made to preserve her position as nearly as possible until daylight, when, nothing being in sight, she was kept on ENE. under short sail with heavy gales. At about 10 a.m. an altitude of the sun was observed, and the chronometer time noted; but, having run so far without observation, it was plain the latitude by dead reckoning was liable to error and could not be entirely relied upon.

"The longitude by chronometer was determined, by using this uncertain latitude, and it was found to be 15′ E. of the position by dead reckoning; a second latitude was then assumed 10′ north of that by dead reckoning, and toward the danger, giving a position 27 miles ENE. of the former position; a third latitude was assumed 10′ farther north, and still toward the danger, giving a third position ENE. of the second 27 miles. Upon plotting these three positions on the chart, they were seen to be in a straight line, and this line passed through Smalls light.

"It then at once appeared that the observed altitude must have happened at all the three points and at Smalls light and at the ship at the same instant."

A lesser man than Thomas Sumner would have thought this merely an interesting coincidence and let it go at that, without, perhaps, even a log entry. Such a man probably would have floundered on with a wrong notion as to his position until he brought up hard on the beach or blundered onto some sort of a landfall, however uncertain.

Fortunately, for posterity, Sumner did none of these things. He reasoned that although the absolute position of his vessel was still uncertain she must be somewhere on the line he had drawn through the three assumed positions and Smalls Island, an island about 20 miles off Pembroke, Wales. To check this theory he headed the ship ENE. directly along this line and in less than an hour was gratified to observe Smalls Island light, the easterly point of the line, appear practically dead ahead, bearing ENE 1/2 E.

Now, with his position definitely established beyond a shadow of a doubt, Captain Sumner checked back and found that his dead reckoning latitude at the time of his 10 a.m. sight was 8′ too far south which had had the effect of throwing his longitude by Time Sight 31′–30″ too far west. He saw, in an instant, that if he had adopted that position as accurate his ship might well have gotten into a disastrous position.

Captain Sumner saved his ship and the method which he used, the first Line of Position method, served as a starting point for the Marc St. Hilaire Method, published in 1876, the standard of our merchant and naval services down through the days of World War I and several years thereafter. In the "new navigation" of today, there are no less than twenty-nine different line of position methods—eleven American, nine British, two each Brazilian, French, Italian and Japanese and one German and one Portuguese.

Yet all of these methods, including the well known *H.O. 203, H.O. 204, H.O. 208, H.O. 211* and the latest streamliner *H.O. 214,* are essentially the same, based on the solution in different ways of the same astronomical triangle that Sumner saw in his mind's eye. And the originators of these new short methods would be the first to acknowledge gratefully that it all goes back to the skipper of a Yankee windjammer, who on a wintry morning over 100 years ago had the eyes to see, the intelligence to comprehend and the courage to act.

⚓

The "Hamrah's" Unfortunate Voyage

CHARLES F. TILLINGHAST, JR.

ON ONE 8th of June the yacht *Hamrah* left Newport, R. I., bound for Bergen, Norway, in a transatlantic race. She carried a crew of six—Robert Ames, 52, owner and master; Richard Ames, 23, his eldest son, navigator; Henry Ames, 20, his other son; Roger Weed, 23; Sheldon Ware, 20, and myself.

Until the evening of the 8th we were in sight of three of our competitors, *Vagabond*, *Mistress* and *Stoertebeker*, the *Stormy Weather* and *Vamarie* having disappeared by this time ahead of us. We made Noman's Land and there tacked to the south and later to the east during the night. About noon of the 9th we were off Sankaty Head, Nantucket, and about 2:30 p.m. dropped this land from sight. It was the last land we saw until we approached Sydney, N. S., on our return.

That evening it began to blow up from east southeast and we stood toward Nova Scotia under a reefed mainsail, full mizzen and headsails. One tack to the south took us near Georges Bank and another toward Nova Scotia, and we found ourselves in the afternoon of the 11th about 30 miles south of Cape Sable, N. S. Here the wind drew to the south and it lightened considerably. We set a balloon jib, full mainsail and mizzen and headed due east. For three days we hardly touched a rope.

Early June 15th the wind went to the northwest and again we were forced to reef the mainsail and carry only one headsail with no mizzen. On the 15th again the wind shifted to the south, but on the 16th went to the northwest and again we reefed. We were now crossing the Grand Banks approximately 160 miles south of Newfoundland. At this point the barometer, which had been falling gradually for two days, began

Reprinted by permission of *The Providence Journal*.

to fall more rapidly, and we began to look for really dirty weather.

The wind shifted to the south on the 17th, thus allowing us to carry full sail, but at noon of the 18th it went to the northeast and began to blow hard. That evening it went to the northwest, and the barometer, having now fallen as low as 28.92, indicated an approaching gale. All that night it blew hard and by morning it had reached gale strength, and had kicked up a high sea. At this time we were doing over 9 knots. The waves from crest to trough were approximately the height of the mainmast and about 60 yards, perhaps 70 yards, from crest to crest.

At the time we were in Lat. 46 N, Long. 40 W, approximately 600 miles east of Cape Race, Newfoundland, although we had not been able to take celestial observations for two days, due to cloudy weather. We were carrying a staysail and double-reefed mainsail on a close reach, and making good weather of it.

Before 9:00 a.m. one or two waves had washed over the *Hamrah's* deck, but they were not serious for anybody on the watch for them and with a good grip. At about 9:00 a.m., however, an unusually large sea broke over her windward or port side. Mr. Ames, the owner, and I were the only ones on deck, Mr. Ames having only just come on deck from below where he had been eating breakfast. I was steering and he sitting beside me on the bottom of the dinghy which was lashed on deck on the port side. The wave went over our heads and when it had passed I saw that Mr. Ames had been washed overboard, even though the life lines and furled mizzen resting in the gallows frame on the stern appeared to afford ample protection.

I yelled for the men below and started to jibe. Before the sail came over, however, Richard, the eldest son, had come on deck, seized the emergency line and jumped from the starboard side. The line was too short for him to reach Mr. Ames, as the vessel was still going away from his father, who was just above the surface, but obviously close to sinking, weighted down as he was with boots and oilskins. Richard let go the rope and swam to his father. The *Hamrah* jibed and I then tried to shoot up beside those in the water in an effort to get as close as possible without too much headway.

We got to within possibly 15 yards of them but then unfortunately as we drew nearer a steep wave pushed them out of reach. A life pre-server was thrown which they got, and a life raft put over which they failed to reach. Having missed them once we paid off on the starboard tack and jibed again. We jibed both times to get to the men as quickly as possible, as without the mizzen and carrying relatively so much sail forward it would have taken much longer to come about and pay off preparatory to shooting into the wind to kill headway.

Mr. Ames was becoming exhausted, and Dick was having trouble to keep himself and father afloat. On the second jibe the main boom broke, therefore making it impossible to bring her about. We immediately put the mizzen on her, but still she would not come about or work to windward. While we were putting the mizzen on her, Henry, the younger son, launched the small rowboat and reached his brother whom he got aboard. His father was by this time possibly drowned; at any rate, I did not see him. The rowboat was to windward, so we jibed again and sailed as close to the wind as possible, hoping that the small boat would be able to back down wind to a point where we could get it. Unfortunately it was swamped at this point. It was then clear that we were going to leeward more quickly than was the swamped small boat.

We worked as quickly as possible to cut away the mainsail which was dragging in the water so that we would be able to beat up to the small boat, but while cutting away the mainsail the gap had widened. With only three of us left on board, we were much handicapped in this work and, indeed, in handling a boat of the size of *Hamrah* at all, under the conditions prevailing.

We tried to keep an eye on the boat, but in the high breaking seas, with the wind blowing a gale, this was difficult. While cutting away the rigging we lost sight of them when they were five or six waves away; perhaps 350 yards. When the mainsail was gone we were able to come about and sailed back and forth for five hours, but were unable to find them. The wind had been increasing all the time so we hove to under the staysail for the next 50 hours, not with any hope of finding the lost men, but because the gale had increased. Under the staysail the *Hamrah* rode comparatively easily.

On June 21st we were able to again get under way. While our nearest port was St. John's, Newfoundland, we decided to sail a more southerly course for Sydney, N. S., in order to avoid icebergs reported numerous to the southeast of Newfoundland, and also because we had a larger scale chart of Nova Scotia.

On the return trip we had moderate to fresh westerly winds all the way with almost continual fog, but no gales, although we were obliged to reef the mizzen once. We were, however, able to obtain satisfactory observations on July 21st, 23rd, 27th and 28th, which enabled me to lay our course without difficulty.

We sailed the 900 miles to Sydney in nine days, arriving there at 10:30 p.m., June 30th, having sighted no vessel to report us.

⚓

48

Long Voyages in Small Boats

C. G. DAVIS

HISTORICALLY, the most interesting of all these transatlantic voyages are those of the three caravals—*Santa Maria, Nina* and *Pinta*—replicas of the originals—that crossed the Atlantic in 1891, and of the small Norwegian craft, a reproduction of an ancient Viking ship unearthed in 1879 at Christiana, that was manned by Norwegians who sailed and rowed this beautifully modeled double-ender across in 1893, just as their forefathers had done centuries ago. The records of some of the earlier long voyages in small boats are hard to get, as such voyages did not seem so much out of the ordinary as now. But about the year 1800, we hear that a Captain Cleveland, of Salem, made a voyage, single-handed, in a fifteen-foot cutter from the Cape of Good Hope to Alaska and the west coast of the United States.

In 1849 J. Miller Cranston, of New Bedford, sailed from that port to San Francisco—13,000 miles—in 226 days in a 41-foot boat named the *Toccao*. That was in the days of the gold fever, when hundreds of Argonauts were going to 'Frisco by way of sailing ships around Cape Horn, or by the short cut across the Panama Isthmus, re-embarking on the west coast.

An engraving showing a small cutter, close-reefed, riding a heavy sea with icebergs to windward, in the August 29, 1863, issue of *Frank Leslie's Illustrated Newspaper,* shows what kind of a craft was the *Skjolfmoen* that was sent out from Bergen, Norway, in command of Rothje Wesenberg and a crew of five men, on the 11th of April, 1863. Loaded with a cargo of herring, codfish and salt, this little 50-ton cutter, 63 feet long, 17 feet beam and 12 feet depth, arrived in Chicago, Ill., in 90 days.

In that same year, 1863, there is a short paragraph in the daily newspapers (as if it were unworthy of more space) announcing that the smallest cutter-sloop, a craft of only 30 tons, had made the passage across the Atlantic from Bordeaux, France, to the West India Islands, sailed by one Joseph Shackford.

On the 17th of June, 1864, there was launched from the foot of New York's Grand Street, East River, a little 15-foot boat, of 4 feet 6 inches beam and 2 feet 10 inches depth, rigged as a brigantine, named the *Vision,* in which, on June 26th, of the same year, Capt. J. C. Donovan and Wm. Spencer, of Providence, R. I., and a dog named Toby, started across the ocean.

"Their only provision for cooking," as an article says, "is a lamp, the voyagers trusting to corned meats for their fare and carrying 55 gallons of water to drink, pure and in coffee. When all was ready Capt. Donovan set his foresail, foretopsail and mainsail, and, giving her the jib, she was headed for Governor's Island on the wind. Standing close in he then tacked ship, reaching over to Whitehall, when he again tacked, heading for Bedloe's Island, and, making one more tack, he stood down the bay with a nice breeze, making at least 8 knots. Their destination was Land's End and London. The pilot boat *Wm. Ball* gave the next tidings of her, having passed her on the 28th of July 45 miles east of Fire Island." I can find no record of her having succeeded in reaching the other side, but I sometimes wonder if a subsequent article referring to the loss, in that same year (1864), of a yawl with a Messrs. Wells and Dawson on board might not refer to the *Vision.*

In 1866 two boats are credited with having crossed the Atlantic Ocean. One was the 27-ton sloop yacht *Alice,* owned by Thomas G. Appleton, who, with two friends, a crew of four men and a steward, left Boston, Mass., at noon on July 12th, 1866, and arrived off the Needles on the evening of July 30th, 1866, where she hove-to till daylight, and the next morning sailed into Cowes Harbor. The *Alice* was 54 feet long, 17 feet 6 inches beam, 6 feet 10 inches draft aft and 2 feet 5 inches forward. She was a full-bodied, roomy, little sloop of the model in general use at that time, with a large cockpit aft of her trunk cabin. She made the return voyage, being 34 days from Cowes to Boston.

The other, and by far the more extraordinary voyage, was the reputed passage across the Atlantic made by the *Red, White and Blue,* a diminutive ship-rigged craft 27 feet in length, 6 feet beam and measuring 2½ tons. She cleared from New York on the 12th of July, 1866, with two men, Messrs. Hudson and Finchly, as crew, and on the 5th of August she spoke the bark *Princess Royal,* of Yarmouth, Nova Scotia. On the 6th and 8th she was nearly lost in a gale by heavy seas, which threw her upon her beam ends; on the 14th of August she made the Bill of Portland, was towed to Margate and finally on to London, the whole voyage occupying 38 days.

Considerable doubt was entertained at the time as to the truth of the story of her passage as told by her log book, but the question was finally decided in her favor. It was claimed by some that she was carried across on the deck of a merchant ship and put overboard at the mouth of the Channel. She afterwards went to France, was towed up the Seine, and was exhibited at the Paris Exposition, where J. D. Jerrold Kelley saw her and writes as follows of her voyage: "I had the privilege both of seeing her and of doubting the story, for which last sin I am now doing penance, by declaring that if the most direct evidence is to be believed, she did sail from New York, did cross the ocean, did speak the Blue Nose bark and did arrive in an English port without assistance. The whole thing was foolish in the extreme, for it proved nothing except to what ends men will go for notoriety."

In 1867 there are two voyages recorded; one the trip of the *American,* a life-raft composed of four cylinders lashed together, which sailed from New York June 4th, 1867, navigated by three men, Capt. John Wikes and Messrs. Miller and Mullane. She arrived at Southampton July 25th. The other was the trip of the dory *Nonpareil,* in which two men, Miller and Lawton, crossed from Gloucester, Mass., to Southampton in 27 days.

The following year, 1868, the attempt was made again, when two men, Messrs. Marshall and French, started across the ocean in the skiff *John T. Ford.* Marshall and the skiff were lost, but French was rescued by a steamer.

Two years later, in 1870, two men and a dog crossed the Atlantic from Liverpool to Boston in 98 days in a little yawl, called the *City of Ragusa.* Frank Leslie's *Illustrated Newspaper* again furnishes us data on this event and an engraving of the boat, a yawl, copied from the *London Graphic,* in which it originally appeared at the time she started from England. Her crew consisted of two, J. C. Buckley, master, and a man named Harper.

"The *City of Ragusa*," says the article, "formerly belonged to the ship *Breeze*, which foundered in a terrific storm in the English Channel. Fourteen of the crew succeeded in reaching Ramsey, Isle of Man, in the long boat, which has thus shown her sea-going qualities. She has since been decked over, with a small cockpit aft. Her cabin has been made as comfortable as the limited space will permit, and she carries three months' provisions and one hundred gallons of water, which can be pumped out of the tanks should it be desirable to lighten the vessel. She has been rigged as a yawl and can also set square sails on both masts, spreading about 70 yards of canvas. In addition to her sails she is fitted with a two-bladed screw propeller, which can be raised when not in use, and is worked by hand in the same manner as a ship's pump. The Captain expects to reach New York in 50 days. The *City of Ragusa* is the smallest vessel ever cleared at the Liverpool Custom House, and, curiously enough, the gentleman who performed that duty for her also cleared the *Great Eastern*, the largest."

We sometimes run across information in unexpected places, and that is the case with the next entry. On page 102 of Hayden's *Dictionary of Dates* there is an entry which reads: "Boat Voyage. Alfred Johnson, a young man, started from America (Gloucester, Mass.) in the *Centennial*, a dory 20 feet long, sloop-rigged, on June 15th, 1876, and landed at Abercastle, Pembrokeshire, August 11th, 1876."

In 1877 Thomas Crapo and his wife, Joanna, of New Bedford, Mass., sailed in a 20-foot, sharpie-rigged decked whale-boat named the *New Bedford*, from New Bedford to Penzance, England, in 49 days. The *New Bedford* was built by Samuel Mitchell on Fish Island, connected with the city at that time by a foot bridge. She was 20 feet long, 6 feet 2 inches beam, 33 inches deep and drew 13 inches of water. She had two small deck hatches each 18 x 24 inches. She carried two leg-o'-mutton sails, the foresail of 15 square yards, the mainsail of 10 square yards.

They left New Bedford May 28th, 1877, amid cheers, ran across to Vineyard Haven, where they stayed overnight, then sailed the next day across the bay to Chatham. Here they stayed several days while changes in the deck hatches were made and 200 pounds of weight was added to her keel.

It was June 2nd when they finally started for Europe. Many times they were forced to ride to sea anchors. Only once did they leave their own craft, and that was to pay a visit aboard the bark *Amphitrite* and have a hot dinner and a chance to stretch their limbs. Once they were nearly run down when they both fell asleep at the same time, but at 4:00 p.m. on July 21st, the *New Bedford* safely passed Land's End, being saluted by the lighthouse keeper, and tying up alongside a fishing boat in Penzance Harbor at 11:00 p.m. Both of the crew were so tired that they fell asleep while waiting for the water to boil to brew a pot of coffee.

Capt. W. A. Andrews and Josiah W. Lawlor were very much in the public eye when I became old enough to take an interest in affairs nautical. It was this Capt. Andrews and his brother who sailed from City Point, Boston, Mass., on Monday, June 3rd, 1878, for France in the *Nautilus,* the smallest boat yet to attempt this venture. Built by Higgins & Gifford, of Gloucester, this dory was 19 feet long over all; 15 feet 6 inches on the bottom and 27 inches deep. She had an 18-foot mast, and a yard 23 feet long, spreading a triangular lateen sail of 25 square yards. She had aboard two other sails; a square sail of 5½ yards and a tiny storm trysail of 1½ yards of stouter canvas. The adventurers had no idea of navigation, but spoke many vessels, and so kept on their right course, arriving in 45 days at Mullion Cove, Cornwall, England.

Three years later, in 1881, the seaworthiness of a ballenera, or Buenos Ayres whaleboat, was proved by the safe passage made by two Italians from Buenos Ayres to Caprera on the Mediterranean Sea in one of these small boats.

Benjamin Radford and Charles Moore, in July, 1881, fitted up a ship's lifeboat 19 feet long on the keel, 6 feet beam and with a draft of 13 inches, in which they expected to cross the ocean in 35 days. She was rigged as a brigantine and named the *William R. Grace,* in honor of New York's mayor. Both men had followed the sea for many years. Besides a swinging rudder for calm weather they had a long steering oar for rough water work; and while the hull was open, their food was kept dry in lockers built under the seats. There was a sketch of her in an illustrated magazine of July, 1881, but no record is found of their having made the trip.

Seven years later Capt. W. A. Andrews, of the *Nautilus,* built another boat only 15 feet long, 5 feet beam, 2 feet 3 inches deep. He first named her the *Mermaid,* but a showman offered him $500 down and a tour at $100 a week to change her name, so he called her the *Dark Secret,* after a play of that name then touring the States. He left Boston on the 17th of June, 1888, and after being buffeted about for 68 days in terrible weather he was picked up by the ship *Nor* only 150 miles from Boston and taken back to America.

Capt. Joshua Slocum, who later achieved fame as skipper of the *Spray,* was captain of the ship *Aquidneck,* which had been wrecked on a sandbar off the Brazilian coast. From the material in the wreck he built himself a big, decked dory, 33 feet long, 7 feet 6 inches beam and 3 feet deep, rigged with three masts and batten sails like a Chinese junk. On July 23rd, 1888, this queer craft, which he had named the *Liberdade,* with the Captain, his wife, baby and two sons, started on a 5,000-mile voyage for his home in the United States.

In 1891 the newspapers had long accounts of two dories that raced across the Atlantic. They were each 15 feet long. Capt. W. A. Andrews owned one of the boats, called the *Mermaid,* and J. W. Lawlor the other, which he named the *Sea Serpent.* They left Boston at the same time on June 17th. The *Mermaid* had good weather for 35 days, when a gale sprang up and she was half swamped several times, was capsized on August 17th, according to the Captain's story, and finally, on August 22nd, with some 600 miles still to go, he abandoned the trip and was taken, boat and all, to Antwerp on the steamship *Elbruz,* where he sold his boat and returned to America.

The *Sea Serpent* also was capsized, but with the 315 pounds of lead on her keel and the Captain's weight, she was righted again; this was on July 10th, and then for 23 hours she lay to a sea anchor. On the 19th she spoke a German bark from Plymouth, England, bound for Quebec, Canada, and on July 30th arrived at Land's End, England.

In spite of his previous hard luck, Capt. Andrews again essayed to cross the Atlantic in July, 1892, in a smaller boat than he had ever before tried. Capt. Andrews at this time was a tall, well-built man 49 years of age, a piano-maker by trade and a man who courted notoriety.

The *Flying Dutchman,* as he called this latest boat, was to sail from Atlantic City, N. J., where he had built her, to Palos, Spain, in about fifty days, as he estimated. She was a double-ended craft, somewhat like a sneak-box, 14 feet 6 inches in length and 5 feet 5 inches in width, with a depth of nearly three feet. The sides, bottom and deck were of ½-inch cedar, covered with canvas, then oiled and painted. A small two-foot by three-foot cockpit was built in, water-tight, with dry-stowage below decks. She was not sub-divided into airtight compartments, as one might expect with a keel of 300 pounds of lead under her. Her rig was a sort of sliding gunter mainsail, with a jib. The mast was 8 feet above deck and the boom 13 feet long with a 10-foot gaff, the total spread being 15 square yards. There was some delay about the start, but he eventually left Atlantic City, and that was the last ever seen of Capt. Wm. A. Andrews or the *Flying Dutchman*.

By far the most remarkable voyage in a small boat is that made by Capt. Joshua Slocum, single-handed, around the world in the yawl *Spray,* a Connecticut oyster smack, 37 feet long, 14 feet 2 inches wide and 4 feet 2 inches deep.

The year following *Spray*'s departure, in 1896, two young men, George Harbo and Frank Samuelson, rowed in a lightly built, double-ended, lap-strake rowboat called the *Fox* from New York to Havre, France. They left the Battery, New York Harbor, on June 6th, and 55 days later landed at the Scilly Isles, where they rested a couple of days, and then rowed on, arriving at Havre August 7th. They were capsized once, but the air tanks in the ends of the boat kept her up and helped right her, and as each man had a rope about his waist, they regained the boat, bailed her out and went on. On July 1st the schooner *Leader* offered to take them back, but they declined and rowed on. On July 10th, while riding to a sea anchor in a gale off the banks of Newfoundland, they were capsized. On July 15th the Norwegian bark *Cito,* bound from Quebec to Pembroke, picked them up and gave them the first warm food they had had since the capsize, when much of their outfit was lost. The Norwegian bark *Eugene,* from Halifax for Swansea, next sighted them and supplied them with bread and water, and on August 1st land, the Scilly Isles, was sighted and reached.

The next small boat to cross was the sloop *Great Western* in 1899, commanded by Captain Howard Blackburn, who, alone, sailed from Gloucester, Mass., to Gloucester, England, in 62 days. A remarkable thing about this was that Blackburn was without hands and feet, as they had been frozen and amputated as the result of a winter fishing trip he had made. In 1901 Captain Blackburn again crossed the Atlantic alone, this time in the 25-foot sloop *Great Republic,* sailing from Gloucester, Mass., to Lisbon, Portugal, in 39 days.

Almost as remarkable as the voyage of the *Spray* was that of the *Tilikum,* a rebuilt Alaskan war canoe 40 feet long, 5 feet 6 inches beam and 18 inches draft, 10 inches of which was a false keel fitted to her 500 pounds of ballast, Captain J. C. Voss, a Canadian, undertook this trip on a wager of $5,000 that he could beat Captain Slocum's achievement. This unique craft was decked over, a small cabin and cockpit were built amidships, and she was rigged with three small masts as a fore-and-aft schooner with one jib and leg-o'-mutton spanker.

The *Tilikum* sailed from Victoria, B. C., on May 22nd, 1901, with Capt. Voss and Mr. Luxton, who had started the wager, on board. She visited the South Sea Islands, Australia, New Zealand, Torres Strait, Arafura Sea, Indian Ocean, Cape of Good Hope, went across the South Atlantic to Pernambuco, then by way of the Azores to England, where, after being exhibited, she was sold. Her cruise occupied three years three months and twelve days, and it is calculated she sailed about 40,000 miles.

In 1902 a queer, flat-sided, flat-bottomed yawl, the *Nina,* sailed by a man named Frietsch, tied up in the basin at the Battery, New York, having come by way of the inland waterways and canals from Milwaukee, Wis. I looked her over one noon-time, and of all the rough-looking jobs of boat-building she was the worst piece of patchwork I think I ever saw. Butts were so numerous in her side planks as to make one think that she was built of pieces of packing boxes. Her gear was of the poorest and yet, after being stocked up by donations from spectators and given a free tow down the bay, she did manage to arrive on the other side, where, I believe, she went ashore and was wrecked.

It was this same year that the *Abiel Abbott Low,* the first motorboat to cross the ocean, was navigated across by Captain W. C. Newman, then forty years old, and his sixteen-year-old son, C. E. Newman, of College Point, L. I. The *Low* was a trunk-cabin launch 38 feet long, 9 feet beam, drew 3 feet 8 inches of water and had 2 feet 3 inches least freeboard when she started with 800 gallons of kerosene fuel for the 10 horsepower motor, 250 gallons of fresh water and food for 60 days. She left New York July 9th, 1902, and on August 14 arrived at Falmouth, England, having consumed about 440 gallons of fuel. She continued on in easy runs to London, where she tied up in the West India Dock September 20th.

In 1903 Captain Ludwig Eisenbram is credited with sailing the *Columbia II,* a little ship only 19 feet long and 6 feet wide, from Boston to Gibraltar in 101 days. How authentic this is I do not know, but a notice in the November 21st, 1903, issue of the New York *Times* announces his arrival at Gibraltar on November 20 and says he left Boston, Mass., on August 11th of that year.

By far the queerest voyage of all those made across the Western Ocean was that of the *Vraad,* a lifeboat designed by Captain Abe Brudel and shaped like an English walnut. She was 18 feet long, 8 feet wide and 8 feet deep, built of ⅛-inch boiler plate iron riveted to circular iron frames. She drew 4 feet of water and a heavy wooden buffer at the water line acted as a bilge-keel to stop her rolling. Above water one portlight on each side forward admitted light and air, and she was navigated from below in stormy weather. A hollow steel mast 20 feet long, stepped away forward, raked aft, and carried a lug sail of 250 square feet.

Captain Brudel, then twenty-three years of age, started from Aalesund, Norway, on June 27th, 1904, with three companions. They had fine weather crossing the ocean, but gale after gale made it wet sailing from the Bay of Fundy down the New England coast. They were three times blown out to sea by screaming nor'westers, but, their boat being fitted with a centerboard, they were able to beat back each time, and finally landed high and dry on the beach at Gloucester on January 7th, 1905, having mistaken the lights along the shore for the anchor lights on vessels in the harbor.

Two days before the *Vraad* ended her unique trip a long, narrow, torpedo boat, named the *Gregory,* driven by gasolene engines, left Perth Amboy, N. J., bound for Sebastopol, Russia, which port she eventually reached. She put back once, went into Bermuda for a harbor, stopped at Punta Delgada in the Azores for gasolene, and then went on to Sebastopol through the Mediterranean and Black Seas.

An unfortunate ending in the last stage of her voyage blotted out the yawl *Pandora* on her trip across from New York to England after coming around the Horn, all the way from Bunbury, West Australia, which she left May 3rd, 1910.

Built very much on the same model as the famous *Spray,* the *Pandora* was 37 feet long, 14 feet beam and drew 4 feet of water. Captain George Blythe, of Coventry, England, the owner, and Captain Peter Arapakis, her navigators, had some adventures to tell when the *Pandora* anchored in Gravesend Bay on June 23rd, 1911. They had touched at Melbourne on May 29th, 1910; Sydney, Australia, on August 16th;

Auckland, New Zealand, September 4th, where repairs kept them until October 2nd, when they sailed for Pitcairn Island. On the 21st of November this land was left astern, and December 12th Easter Island was reached. On January 16th, 1911, the little vessel passed Cape Horn, and in a nasty gale on January 22nd, while both men were below, *Pandora* was rolled completely over by a huge sea, came up dismasted, and was towed into the West Falkland Islands by a whaler, where repairs were made. On March 4th she left for St. Helena, leaving there April 26th for Ascension Island, where she arrived May 3rd, 1911, and from there she made the run to New York, only to reach the port of missing ships when the end was so near, as she never was reported after clearing Sandy Hook, bound for England.

Three attempts to cross the Atlantic were made in the year 1911. Two boats were wrecked, and one, the smallest, crossed safely. The sloop *Theresa,* built at New Bedford in 1864, 45 feet long, 17.7 feet beam and 4.5 feet draft, with a 20-horsepower gasolene motor, sailed from Narragansett Bay for the Cape Verde Islands with Captain Joaquin Rene, Jose Fonseca and Manuel Andrade aboard, but she never arrived. She was so old and rotten she split in two and sank, and a steamer rescued her crew.

A 50-foot motorboat, named the *Romania,* 12 feet wide and 5 feet deep, fitted with a 37-horsepower gasolene motor, and commanded by Captain John Weller, who built her at Carlstadt, N. J., started from Bridgeport, Conn., July 19th. She was completely wrecked on Chelbogue Point, Yarmouth, Nova Scotia, on July 25th, but her crew was saved.

The smallest boat of the three, the little 25-foot skip-jack yawl *Sea Bird,* with Captain Thomas Fleming Day, of New York, and Fred B. Thurber and Theodore R. Goodwin, of Providence, R. I., made a quick and safe passage across in June, 1911. *Sea Bird* was fitted with a single-cylinder 3-horsepower motor. She was 8 feet 4 inches beam, 3 feet 8 inches draft and carried 400 feet of sail.

..*Detroit,* the 35-foot, double-ended, motor lifeboat, which crossed the Atlantic in 1912 with Captain T. F. Day, Wm. Moreton, Charles E. Earle, Jr., and Wm. Newstead aboard, was built at Port Clinton, Ohio, for Com. Wm. E. Scripps, of Detroit, Mich. She was 10 feet beam, had a draft of 5 feet when loaded and a displacement of 14 tons. The entire midsection, under her deck, was filled with big steel tanks which, with two tanks on deck, carried a total of 960 gallons of gasoline. The crew lived down forward and the engine room was away aft, with a raised hatchway to keep the water out.

Launched on June 25th, 1912, *Detroit* arrived in New York by way of the Great Lakes and Erie Canal on July 12th and fitted out at College Point, L. I. On July 14th she left the New Rochelle Y. C., but stopped at Cottage City, Martha's Vineyard, to fill her tanks and let the skipper say good-bye to his family, who were summering there. Leaving Vineyard Haven July 16th, *Detroit* arrived in Queenstown, Ireland, on August 7th, her 16-horsepower, 2-cylinder motor driving her about six miles an hour. She reached St. Petersburg September 12th.

Among the many unsubstantiated voyages in small boats may be mentioned that of Captain James Cassels, of Glasgow, and John C. Maloney, who claimed to have crossed the Atlantic in 97 days in a 32-foot auxiliary sailboat, named the *Imp,* which arrived at Halifax on June 25th, 1914, an account of which appeared in the New York *Herald* in July, 1914. They claimed to be on their way to San Francisco by way of the Panama Canal. The *Imp* was said to have left Dublin, Ireland, on March 14th. Rough weather was experienced from the time she left Queenstown, but she not once used her motor.

In 1920, W. W. Nutting sailed over in the *Typhoon,* making a fast passage of 22 days, from Baddeck to Cowes. He returned the same year, arriving in New York in November after a rough passage.

⚓

49

The Case of the Bare Behind

FREDERIC M. GARDINER

IT WAS between Cape Ann and Minot's that I had my most amazing nautical experience, and I am going to tell it as an instance of the strange sort of thing which can happen at sea. It was not behind any islands, but in the Atlantic Ocean five miles off Nahant. We were sailing over from Scituate to Marblehead; there was a light breeze with a smooth sea; dusk was falling and there were no craft in sight except a little power cruiser off to leeward going about her own business.

One of the crew thought he heard a hail, and we all listened. It came again and, peering to weather, we made out an object bobbing in the water which proved to be a *man's head*. Without asking questions we beat our way up to it and hove to alongside. The man in the water —chubby and appearing perfectly calm and contented—refused to be taken on board before we answered a question.

"Are there," he demanded, "any women on board?"

That was a new one. We reassured him, and dragged him over the side. He was stark naked! We wrapped him up, put a hot toddy in his hand, and plied him with questions. What had occurred turned out to be perfectly simple, and the sort of thing which might happen to anyone. It was a hot day; he had taken out his motor cruiser—by himself—to cool off. For further comfort he had stripped off all his clothes. Then he noticed that the painter had slipped off the fo'c'sle head and was trailing in the water. He went forward on the slippery deck, the boat took a lurch, and over he went—while his boat ran merrily on.

Not to prolong the story, it turned out that the little cruiser we had seen barging about was his. After a rather exciting chase we managed to catch up with her and put him back on board. The last I saw of him was his broad, bare behind disappearing below into his cabin, and that is the last I have *ever* seen or heard of him.

It only goes to prove that you never can tell what entertaining adventure may be awaiting you when you put to sea on a cruise.

⚓